ECOPHILOSOPHY IN A WORLD OF CRISIS

Building on its origins at a seminar in Oslo organized by two of the editors, this book combines classic texts of Nordic ecophilosophy and the original contributions of those influenced by this tradition to present the view that critical realism is indeed a worthy intellectual tradition to carry forward and further develop the work of the founders of Nordic ecophilosophy.

It was clear at the seminar that there was a promising convergence of interests and themes in the two approaches; while at the same time, within the Nordic ecophilosophical tradition, there was appreciation of the capacity of critical realism, with its provision of a robust philosophical ontology and generation of totalizing immanent critiques of Western philosophy, to provide an expansive and secure home for the development of ecophilosophical work generally.

If there is a single overarching theme of critical realist philosophy, it surely must be that of the unity of theory and practice, which Bhaskar, following Hegel, has also called 'seriousness'. This makes the applicability, relevance and actionability of critical realism key considerations for critical realists. There can be no doubt that this concern was shared fully by the Nordic ecophilosophers; and this quality of 'seriousness' is a striking feature of the Nordic contributions presented in this book.

Roy Bhaskar is the originator of the philosophy of critical realism, and the author of many acclaimed and influential works including *A Realist Theory of Science, The Possibility of Naturalism, Scientific Realism and Human Emancipation, Dialectic: The Pulse of Freedom, Plato Etc., From Science to Emancipation* and (with Mervyn Hartwig) *The Formation of Critical Realism*. He is an editor of *Critical Realism: Essential Readings* and *Interdisciplinarity and Climate Change* and was the founding chair of the Centre for Critical Realism. He is currently a World Scholar at the University of London Institute of Education.

Karl Georg Høyer is Professor and Research Director at Oslo University College. He holds a Master's Degree in technology and a PhD in social sciences with a dissertation on 'Sustainable Mobility'. Most of Høyer's research is related to sustainable development, with a main focus on transport and energy.

Petter Næss is Professor in Urban Planning at Aalborg University, Denmark. His main research interests include: land use and travel; impacts and driving forces of urban development; and the philosophy of science. His most recent book is *Urban Structure Matters* (Routledge, 2006).

Ontological Explorations:

Other titles in this series:

From One 'Empire' to the Next
Radha D'Souza

Science for Humanism
The recovery of human agency
Charles R. Varela

Philosophical Problems of Sustainability
Taking sustainability forward with a critical realist approach
Jenneth Parker

Dialectic and Difference
Dialectical critical realism and the grounds of justice
Alan Norrie

Interdisciplinarity and Climate Change
Transforming knowledge and practice for our global future
Edited by Roy Bhaskar, Cheryl Frank, Karl Georg Høyer, Petter Næss and Jenneth Parker

Conversations about Reflexivity
Edited by Margaret S. Archer

Relational Sociology
A new paradigm for the social sciences
Pierpaolo Donati

Sociological Realism
Edited by Andrea M. Maccarini, Emmanuele Morandi and Riccardo Prandini

The Economics of Science: A Critical Realist Overview
Volume 1: illustrations and philosophical preliminaries
David Tyfield

The Economics of Science: A Critical Realist Overview
Volume 2: towards a synthesis of political economy and science and technology studies
David Tyfield

Ecophilosophy in a World of Crisis: Critical Realism and the Nordic Contributions
Edited by Roy Bhaskar, Karl Georg Høyer and Petter Næss

For Cheryl Frank
1946–2010

CONTENTS

PREFACE

Roy Bhaskar

This book had its origins in a two-day seminar organized in November 2009 by two of the editors, Karl Georg Høyer and Petter Næss. Three of the other contributors – Sigmund Kvaløy Setereng, Arne Johan Vetlesen and Roy Bhaskar – were also present and spoke at the seminar. This meeting was followed by a planning meeting of the Nordic Network of Critical Realism, to which the Nordic ecophilosophers were invited. At the end of the seminar it is probably true to say that most of the ecophilosophers would have agreed with what seemed to be Sigmund's view, namely that critical realism was indeed a worthy intellectual tradition to carry forward and further develop the work of the founders of the Nordic ecophilosophical tradition. Certainly it was clear that there was a promising convergence of interests and themes in the two approaches; while at the same time, within the Nordic ecophilosophical tradition, one could discern some appreciation of the capacity ot critical realism, with its development of a robust philosophical ontology and totalizing immanent critiques of Western philosophy, to provide an expansive and secure home for the development of ecophilosophical work generally.

This book follows on the trail of an earlier volume, *Interdisciplinarity and Climate Change*,[1] and that volume in turn followed up on the promise of the Tromsø International Association of Critical Realism Conference in August 2006. The ground for this had itself been laid by my period there as an Adjunct Professor of Philosophy at the Norwegian National Peace Centre in 2004–6. I remain very grateful for the generous hospitality shown to me by my Norwegian hosts, in Tromsø, as in Oslo, Lillehammer and elsewhere in Norway. For me, Tromsø had much in common with the Himalayas. In both places I experienced a palpable, and eerily similar, feeling of height, of being, as it were, on top of the world; in both places also there seemed to be a wonderful simplicity and innocence of nature, and the local environment, both internal and external, non-human and human.

These trips were in turn embedded within a longer span of time spent largely, if not exclusively, in Scandinavia, beginning with my sojourn as a Fellow of the Swedish Collegium for Advanced Studies in the Social Sciences in Uppsala over 2003–4, and including a memorable spell as a Guest Professor in the Swedish Institute of Disability Research at the University of Örebro. In this period, immediately following the publication of my books on metaReality in 2002, I became increasingly concerned with the problems of applying critical realism, in its three phases (of basic critical realism, dialectical critical realism and the philosophy of metaReality), to the problems of the world: and in particular to the resolution of the problem of alienation, and the crises with which it was associated, on all four planes of social being, namely, material transactions with nature, social interactions between people, social structure and the stratification of the embodied personality.

Now if there is a single overarching theme of critical realist philosophy, it surely must be something that follows from our capacity for reflexivity, the theme of the unity of theory and practice, which following Hegel, I have also called 'seriousness'. This makes the applicability, relevance and actionability of critical realism key considerations for critical realists. There can be no doubt that this concern was shared fully by the Nordic ecophilosophers; and this quality of 'seriousness' is a striking feature of the Nordic contributions presented in this book.

During this time in the Nordic world my constant companion was Cheryl Frank, who I had met in Ithaca, New York in November 2002. She died in January 2010, and I am grateful to my co-editors for agreeing to dedicate this book to her memory.

Roy Bhaskar, April 2011

Note

1 Edited by Roy Bhaskar, Cheryl Frank, Karl Georg Høyer, Petter Næss and Jenneth Parker, Routledge 2010.

ACKNOWLEDGEMENTS

We would like to thank Britt and Eva Hägerstrand, Brit Olofsdottir from the FORMAS research council and Gösta Carlestam, editor of previous Swedish-language books by Torsten Hägerstrand for permission to reproduce the chapters by Torsten Hägerstrand.

We are similarly very grateful to Elisabeth and Benedict von Wright for permission to reproduce the chapter by Georg Henrik von Wright. We are also immensely grateful to Sigmund Kvaløy who has allowed us to reproduce his article on Servoglobe, and also to Jon Wetlesen who has given us the right to abbreviate and translate his article from Norwegian to English.

Finally, we owe a great debt of thanks to Chris Saunders for translating the texts in this book by Peter Wessel Zapffe, Jon Wetlesen, Georg Henrik von Wright and Torsten Hägerstrand from their original Norwegian and Swedish-language versions into English.

RRB, KGH and PN,
May 2011

1

INTRODUCTORY PERSPECTIVES

Karl Georg Høyer and Petter Næss

'Surely I am a critical realist now', said the Norwegian ecophilosopher Sigmund Kvaløy Setereng at a seminar organized in the autumn of 2009 on similarities and differences between the Nordic ecophilosophical tradition and the philosophy of critical realism. We think his now departed ecophilosopher colleagues Arne Næss, Georg Henrik von Wright and Torsten Hägerstrand would have agreed if they had the opportunity, and certainly many of the other more or less active supporters and contributors to the Nordic ecophilosophical movement.

Yet, except from Sigmund Kvaløy Setereng's recent exclamation of affiliation, none of the leading figures of the Nordic ecophilosophy literature have ever, according to our knowledge, referred to the philosophy of critical realism in any of their writings or speeches. Conversely, most of the leading critical realist authors seem to have been unaware of the Nordic ecophilosophical tradition until a few years ago.[1] This is a bit strange, since most of the critical realist literature and the literature within Nordic ecophilosophy has been produced during the same period of time, namely since the early and mid-1970s (with Peter Wessel Zapffe's much earlier ecophilopsophical texts as an exception). However, there are indeed several affinities and overlapping perspectives. The purpose of this book is to explore some of these arenas. In particular, we hope that the book will be a contribution to the intellectual basis for responsible, solidary and emancipatory actions in our rapidly changing world, where nature as well as the conditions for human welfare are in profound crisis.

There are indeed several arenas of common ground between critical realism and Nordic ecophilosophy. This applies to the part of critical realism often referred to as 'basic' (particularly developed by Bhaskar in *A Realist Theory of Science* (1975) and *The Possibility of Naturalism* (1978) and in later elaborations and applications by Bhaskar and other authors) as well as to dialectical critical realism and the philosophy of metaReality. Below, seven main arenas of common ground between Nordic ecophilosophy and critical realism will be discussed:

- critical realist ontology
- the world as differentiated, stratified and consisting of open systems
- interdisciplinarity
- structure and agency
- non-dualism between humans and nature
- fact, value and social critique
- metaReality and non-anthropocentrism.

Critical realist ontology

Critical realists and the Nordic ecophilosophers share a clear realist ontology. Both reject conflations of what there is with what we know. Instead, both the ecophilosophers and critical realists emphasize the fallibility of our knowledge – a position logically implying that there is more to the world than what we (believe we) know. Critical realism as well as the Nordic ecophilosophers express a strong anti-positivism and were to high extent developed as fundamental critiques of the then (in the late 1960s and early 1970s) dominant positivist paradigm. A common critique shared by ecophilosophers and critical realists is that positivism ignores underlying, non-actualized structures and the open systems and multi-mechanism situations characterizing societies as well as nature. Both critical realists and the Nordic ecophilosophers thus reject the empiricist limitation of the notion of knowledge to that which can be documented through exact measurement and quantification. Moreover, Arne Næss explicitly dissociated himself from the Cartesian denial of the existence of colour, taste and smell in nature as real entities. Both Næss and Kvaløy, as well as Hägerstrand, have on several occasions given very fundamental and explicit critiques of positivism in science.

Generally, the Nordic ecophilosophical literature must be characterized – as critical realism – as being in clear opposition to irrealism and radical social constructionism. The notion that 'there does not exist anything beyond the text' is as utterly strange to Nordic ecophilosophers as it is to critical realists. Although the Nordic ecophilosophers' writings in the 1970s usually did not address the postmodern non-realist positions explicitly, they implicitly clearly rejected them. Kvaløy and Hägerstrand were perhaps the ones most explicitly positioning themselves in opposition to the individualism and voluntarism that came to characterize much postmodern discourse. In later contributions, not least in Arne Johan Vetlesen's recent writings, the criticism against postmodern relativism and nihilism is stark and highly consistent with critical realist critiques.

A differentiated, stratified and open-systemic world

The critical realist view of the world as a predominantly open system characterized by the simultaneous operation of a multitude of different causal mechanisms resonates with the opinion of the Nordic ecophilosophers that 'everything is related to everything'. For example, in clear contrast to the Humean conception of causality, Hägerstrand (1974/1991, p. 169) points to the necessity of being able to pinpoint the reasons for 'non-events' by tracing the barriers preventing certain types of events and states from occurring. This is, according to Hägerstrand, probably necessary in order to 'understand in a deeper sense the nature of ecological conditions common to both living beings and the man-made world of artifacts'.

A criticism of the classical Western worldview revolving around the concepts of regularity, predictability and control is also manifest in the writings of Georg Henrik von Wright. Compared to Hägerstrand, he tends to emphasize the chaotic elements in the world to a higher extent. He also seems to be more reluctant to accept the idea of cumulative knowledge than the other Nordic ecophilosophers. The influence of chaos theory and Kuhnian paradigm theory on von Wright's ontology should however not be exaggerated.

Interdisciplinarity

With its ontology of reality as differentiated, stratified and open-systemic, critical realism understands empirical phenomena as being the results of a multitude of simultaneously working causal mechanisms. The impacts of one single causal factor are also often multiple and complex. It follows logically from this ontology that empirical phenomena can seldom be explained only by causal factors 'belonging to' one single discipline. The same applies to the impacts of one particular phenomenon. An interdisciplinary approach is therefore necessary in order to avoid drawing false inferences about the causes of an observed event as well as to assess the impacts of, for example, a particular technical solution. According to critical realism, perspectives of different disciplines must be acknowledged as addressing different levels or aspects of reality instead of being considered as mutually exclusive.

This emphasis on interdisciplinarity resonates very well with the thinking of the Nordic ecophilosophers, and is also explicitly expressed by all of them. Kvaløy (1974) for one stresses the need for generalist insights rather than narrow specialization, and emphasizes the importance of educating what he terms 'super-amateurs', i.e. generalists trained to combine very many different forms of knowledge. Such training was considered a prerequisite for the understanding of ecological complexities and conducive to the respect and love of these complexities. In particular the Norwegian ecophilosophers, Zapffe, Kvaløy and Næss, held that it was necessary to use nature as an arena of learning to study and comprehend ecological complexities. *Friluftsliv* – the simple outdoor life in contact with nature – was therefore pinpointed as an important source of insight.

Hägerstrand also criticized the loss of perspective resulting from the ever-increasing specialization and fragmentation of science. In particular, he was frustrated by the lack of solid analyses of how human, technical and biological systems are interwoven. According to Hägerstrand, a closer cooperation between disciplines would only be the first step. What is needed is liberation from disciplinary straitjackets to make it possible to approach reality in completely new ways (Hägerstrand, 1977/1991). Von Wright, on the other hand, was somewhat more reluctant to embracing the call for holism and synthesis.

While acknowledging the practical need for studying fragments of reality in-depth through the 'glasses' of a separate discipline, Næss (1989) emphasizes the necessity of 'seaming' the fragments together through a more holistic synthesis. According to Næss, this is in particular required each time an eco-political decision is to be made. Although recognizing the need for cross-disciplinary synthesizing in practical decision-making, Næss' view does not seem to incorporate the full implications of an ontology of multi-causality, since he does not explicitly mention the need for each discipline to actively make use of knowledge and perspectives from other disciplines relevant to the phenomenon studied. In the case of the discipline of (neoclassical) economics, Næss still raises a fundamental critique against the narrow and fragmentary analyses, especially when dealing with environmental issues, where both the ecological impacts and their causes are typically looked at in separation from their environmental and social context.

Structure and agency

Critical realism's transformational model of the connection between social structure and agency (Bhaskar, 1979 and 1993; Archer, 2000), where structures and agents are mutually influencing each other but with a time-lag due to which our actions in a given moment are always constrained and enabled by the structural conditions at hand, fits well with the ways in which the Nordic ecophilopsophers see the agency-structure relationship. Their emphasis is still mostly on the influence of the structural conditions, be they the natural environment, the built environment, the prevailing technologies or the growth-oriented economic system. Hägerstrand's time-geography explicitly points at the many temporal and spatial constraints which in practice limit people's opportunities for choosing what to do, together with whom, and where. Kvaløy together with Zapffe, on the other hand, depict individuals as being to a high extent alienated and trapped by 'false consciousness' lured by, among other things, the 'Disneyland effect' of commercialized leisure opportunities. Næss to a greater extent emphasizes the opportunity and responsibility of individuals in contributing to social change. This is illustrated, among other things, by his appeal to those who agree with the eight 'deep ecological' principles to commit themselves to 'directly or indirectly try[ing] to implement the necessary changes'. We should still not exaggerate these differences between the Nordic ecophilosophers (nor should we exaggerate disparities between critical realist thinkers whose work focuses mainly on either agency or on structure). After all, it was the 'structuralist' Kvaløy who broke from his urban life as a university scholar to pursue a different individual lifestyle as a mountain farmer in a remote rural region.

Non-dualism between humans and nature

Common to critical realism and the Nordic ecophilosophers is the conception of humans as part of nature and as highly dependent on nature for our survival and flourishing. More generally, critical realism and the Nordic ecophilosophy share an understanding of the world as consisting of open systems, with interdependencies across scales as well as levels of reality. The critical realist conception of reality as stratified and differentiated, consisting of different 'layers' where new causal powers and properties that are irreducible to lower layers emerge when moving from a lower to a higher layer, where social life by necessity involves the powers and mechanisms of, among others, physics, chemistry and biology but cannot be reduced to the powers and mechanisms of lower strata, squares well with the ecophilosophical conceptions of non-duality between the human species and the rest of nature.

Needless to say, human dependency on nature is strongly emphasized in the Nordic ecophilosophy. Apart from the need for natural resources for nutrition and material products, our dependency on nature involves nature's positive contribution to joy, health and learning (emphasized in particular in Næss's embracing of *friluftsliv* – the simple outdoor life in co-presence with nature – as a practice potentially subversive to the streamlining of consumerism). There is also an opposite side of the coin: the alienating 'pipeline environment' where only those aspects of human nature compatible with productivism in the service of industrial growth society are allowed to unfold. The rootedness of society and humans in nature is also clearly expressed in critical realism: the 'lower strata' are always with us, and 'material transactions with nature' are one of the constitutive parts of 'the four-planar social being' discussed in Bhaskar's (1993) dialectical critical realism.

In line with the strong non-anthropocentrism of Nordic ecophilosophy, critical realism goes against 'the anthropic fallacy', which, according to Bhaskar is 'the exegesis of being in terms of human being'. Critical realism's rejection of anthropocentrism applies to ontological as well as ethical anthropocentrism. The 'sixth level of ontology' referred to in the philosophy of metaReality involves a re-enchantment of the world, including acknowledgment that all species have value in themselves, irrespectively of their utility for humans. Needless to say, this is consistent with the non-anthropocentric and ecocentric views of the Nordic ecophilosophers. Moreover, critical realism's rejection of cognitive triumphalism (Bhaskar, 1993, 2002) is in line with the underlying ideas of the precautionary principle, which has been strongly supported by the Nordic ecophilosophers as a guideline for environmental policy.

Fact, value and social critique

Both critical realism and the Nordic ecophilosophical tradition consider that values can be rationally justified, i.e. fact-value dualism is rejected. Whereas Næss and other Nordic ecophilosophers stressed that the science of ecology cannot itself prescribe how humans ought to act in relation to the natural environment, they also rejected the notion that values and norms cannot be subject to rational justification or critique. Næss contended that many of the norms guiding social practices and individual behaviour are based on some tacit normative and descriptive premises.

Sometimes, the descriptive premises are false, and the practice-level norms should then be changed in order to conform to more fundamental values. Often, however, there is a lack of reflection about the consistency of practice-guiding norms and higher-level values, resulting in a situation of 'norm pyramids with hollow middle parts', as metaphorically depicted by Næss (1989). Critical realist arguments, put forward by Bhaskar (1998) in *The Possibility of Naturalism* and later developed further by himself as well as by Alan Norrie (2010) and Andrew Sayer (2000 and 2011), are much in line with the above lines of thought. For example, taking the satisfaction of some basic human need (like nutrition) as a supposedly widely agreed-on normative premise, knowledge about conditions causing starvation can be used to underpin norms of emancipating people from such conditions (Bhaskar, 1998).

Moreover, critical realism as well as the Nordic ecophilosophy have functioned as platforms – or as underlabourers, to use a relevant critical realist term – for critique of 'industrial growth society' with its technocracy and monodisciplinary 'tunnel vision'. In particular, neoclassical economics has been a target of criticism from critical realists as well as from the Nordic ecophilosophers.

MetaReality and non-anthropocentrism

The above similarities between Nordic ecophilosophy and critical realism mainly refer to 'basic' and 'dialectical' critical realism. But there are also clear commonalities between ecophilosophical thinking and the philosophy of metaReality. Næss' notion that when the Self (and the idea of self-realization) is expanded to encompass nature, then hurting nature will be hurting yourself, has much in common with metaReality's emphasis on the shift away from Western dualism to a non-dual model in which emancipation entails 'a breakdown, an overcoming, of the duality and separateness between things' – and also between humans and nature. The metaReality principle of universal solidarity is rooted in a line of argument very similar to Arne Næss's ideas of the expansion of *the Self* displayed in his personal normative system, the *Ecosophy T*. Although not emphasized to the same extent as by Næss, Hägerstrand (1974/1991, p. 168) also expresses a kindred viewpoint, pointing to an enlargement of the concept of community to embrace both living and non-living populations as a possible way to bridge the gap between biological and human ecology.

The similarities here are so striking that one might believe that there must have been some exchange of ideas between Bhaskar and the Nordic ecophilosophers. However, as mentioned above, this does not seem to have been the case. However, Bhaskar's thinking, as well as the ideas of the Nordic ecophilosophers, has arguably been influenced by some important common ideas rooted in Indian and Chinese traditions, for example the Confucian idea of self-cultivation. Bhaskar has clearly drawn on his knowledge of Indian culture and philosophy inherited from his Indian-born father, while Næss and Kvaløy were obviously inspired by the Sherpa cultures they encountered as mountain climbers periodically based in the Rolwaling valley in the Himalayas. Kvaløy and in particular Næss were inspired by Gandhi and made several references to him in their ecophilosophical publications. Kvaløy also draws connections to Hinduism, Taoism and Buddhism. Kvaløy actually later converted to Buddhism and has erected a Buddhist temple on the little Norwegian mountain farm Setereng, where he has lived since the 1970s and from which he has taken his current surname. The direct influence of oriental philosophies on the Nordic ecophilosophers who did not engage in mountain climbing in Nepal is probably lower, but it should be remembered that the period in which many of the main works of Nordic ecophilosophy were written (the early 1970s) was characterized by a zeitgeist in the Western countries where the interest for Indian and Chinese philosophy was high in popular culture as well as in academia.

Note

1 Possible exceptions may include the British critical realists Ted Benton and Kate Soper, but not even in their texts do we find much reference to the Nordic ecophilosophical traditions.

References

Archer, M. S. (2000). *Being Human. The Problem of Agency*. Cambridge: Cambridge University Press.

Bhaskar, R. (1975/2008). *A Realist Theory of Science*. London: Routledge, 2008. (First published by Verso in 1975.)

Bhaskar, R. (1979/1998). *The Possibility of Naturalism: A Philosophical Critique of the Contemporary Human Sciences*. 3rd edn. London and New York: Routledge, 1998. (First published in 1979.)

Bhaskar, R. (1993/2008). *Dialectic: The Pulse of Freedom*. 2nd edn. London and New York: Routledge, 2008. (First published in 1993.)

Bhaskar, R. (2002). *Reflections on Meta-Reality*. New Delhi, Thousand Oaks, CA and London: Sage.

Hägerstrand, T. (1974/1991). 'Ecology under one Perspective.' In Hägerstrand, T. (1991): *Om tidens vidd och tingens ordning – Texter av Torsten Hägerstrand*, pp. 167–170. (On the width of time and the order of things – texts by Torsten Hägerstrand.) Edited by G. Carlestam and B. Sollbe. Stockholm: Statens Råd för Byggnadsforskning. (First published in Bylund, E. et al. (1974): *Ecological Problems of the Circumpolar Area*, pp. 271–276. Luleå: Norrbottens Museum.

Hägerstrand, T. (1977/1991). 'Att skapa sammanhang i människans värld – problemet' (On Making Sense of the Human Environment – The Problem.) In Hägerstrand, T. (1991): *Om tidens vidd och tingens ordning – Texter av Torsten Hägerstrand*, pp. 183–190. (On the width of time and the order of things – texts by Torsten Hägerstrand.) Edited by G. Carlestam and B. Sollbe. Stockholm: Statens Råd för Byggnadsforskning. (First published in 1977 by Kungliga vitterhets-, historie- och antikvitetsakademien (1977): *Konferenser 1*, pp. 5–14. Stockholm: Almqvist & Wiksell.

Hägerstrand, T. (1991). *Om tidens vidd och tingens ordning – Texter av Torsten Hägerstrand*. (On the Width of Time and the Order of Things – texts by Torsten Hägerstrand.) Edited by G. Carlestam and B. Sollbe. Stockholm: Statens Råd för Byggnadsforskning.

Kvaløy, S. (1974). *Økofilosofisk Fragment IV* (Ecophilosophical Fragment IV). Trondheim: Tapir Publishers.

Næss, A. (1989). *Ecology, Community and Lifestyle*. Cambridge: Cambridge University Press.

Norrie, A. (2010). *Dialectic and Difference: Dialectical Critical Realism and the Grounds of Justice*. London: Routledge.

Sayer, A. (2000). *Realism and Social Science*. London: Sage Publications.

Sayer, A. (2011). *Why Things Matter to People: Social Science, Values and Ethical Life*. Cambridge: Cambridge University Press.

2

CRITICAL REALISM IN RESONANCE WITH NORDIC ECOPHILOSOPHY

Roy Bhaskar

On the whole there is a high degree of convergence between critical realism and Nordic ecophilosophy. In this chapter to bring out the resonances, counterpoints, synergies and contrasts between the two, I will look at them from the perspective of the development of critical realism (henceforth normally CR). Ideally, this exercise would be repeated from the vantage point of Nordic ecophilosophy (henceforth NEP). To some extent this is indeed done in this book, in Chapters 1 and 4. However, because of the relative one-sidedness of my perspective in this chapter, let me say clearly at the beginning that there are respects in which NEP has been stronger and better developed than CR. I am thinking in particular of the arenas of space and time, and the levels of social-cultural critique and perhaps popular engagement. Moreover, as perhaps one would expect, Nordic ecophilosophy has been generally much stronger, earlier, deeper and more persistent on ecological matters, while Arne Næss and Sigmund Kvaløy's 'spiritual turns' anticipated mine by a good twenty years or more. However what critical realism has done is to develop, and defend the development, of a philosophical ontology in which it is possible to describe the world, critique mis-descriptions of it (whether in philosophy, science or everyday life) and show why it has been so difficult to do both these things before. Thus in relation to this world which is patently in crisis, critical realism enables us to see more clearly than pre-existing philosophy permitted what needs to be done to improve both our understanding of it, and our practice in it. And, because of this, we can perhaps describe the value of the Nordic contributions more adequately than has hitherto been possible, as well as offering a robust framework for their further development.

In order to identify areas of possible contestation I shall earmark seven touchstone issues:

1 What is the nature of anti-anthropocentricity?
2 In particular, must ethics must be anthropic?

3 What are the grounds for and means of genuine 'interdisciplinarity'?
4 Under what conditions is negentropy possible?
5 And in particular, do ecological considerations dictate an end to growth, and is this the same as the end of development?
6 Is NEP inexorably, even if only tendentially, mono-disciplinary or reductionist?
7 Can one identify self-correcting or homeostatic mechanisms at work in society, and if so how does this affect the nature of critique?

As is customary, I shall differentiate three phases in the development of critical realism:

1 basic critical realism, consisting of transcendental realism, critical naturalism and the theory of explanatory critique;
2 dialectical critical realism; and
3 the philosophy of metaReality.

Section 1: basic critical realism

Transcendental realism

The theme of anti-anthropocentricity is arguably the single most distinctive in critical realism,[1] and resonates with Nordic ecophilosophy from its twentieth-century origins in the tragic pessimism of Peter Zapffe.

There is a double argument at the outset of critical realism. This involves both an argument a) for ontology, and an argument b) for a new ontology. Each of these arguments is in turn duplex. Thus the argument for ontology involves at once an argument for a philosophical as distinct from a scientific ontology, and at the same time an argument for ontology as distinct from epistemology, the conflation of which is critiqued in the concept of the 'epistemic fallacy'. And the argument for the new ontology involves both an argument for the distinctiveness of causal structures, generative mechanisms etc. from patterns of events or empirical regularities (i.e. for the distinctiveness of the domains of the real and the actual [and thence the empirical]) and an argument for the distinctiveness of open from closed systems. These two aspects of the argument for the new ontology are indices of course of the stratification and differentiation of the world.

Now the tradition of NEP, at least implicitly and in practice, generally differentiated ontology from epistemology, even if it did not always so clearly make the distinction. However, for the most part there was little, if any, differentiation between a philosophical and scientific ontology, that is between what the world must be like for certain activities to be possible and the detailed content given to the world by these activities. In general there was not a clearly worked out sense of a characteristic role for philosophy, as distinct from substantive science. For critical realism philosophy concerns itself with just the same world as we know in science and everyday life, but considers it from the point of view of its most abstract and general features, seeking to explicate, via its distinctive (e.g. transcendental) argumentative procedures, precisely what is presupposed by, but not articulated in, our substantive activities.

Similarly, when we turn to the argument for a new ontology, the distinctiveness of the open systems of the world from the closed systems of the laboratory has been well recognized in the NEP tradition. But, on the other hand, the distinction between the objects of scientific knowledge such as structures, mechanisms, etc. and the patterns of events which they generate has not been so decisively registered. In general, the anti-empiricism of NEP has been grounded in recognition of the holistic, rather than stratified, character of reality. Clearly both are necessary.

The most graphic way of representing the anti-anthropocentricity explicit in critical realism is through the conceptualization of what I have called four-planar social being.[2] This revolves around the idea that every social happening or event occurs on at least four planes:

1 the plane of material transactions with nature;
2 the plane of social interactions between people;
3 the plane of social structure *sui generis*; and
4 the plane of the stratification of the embodied personality.

The irreducibilty of the plane of material transactions with nature in the social world means that the material and natural character of what happens is an essential feature of all social explanations. However, the two-way transactions between human beings and nature must be interpreted in the light of an *ecological asymmetry*,[3] weighted in favour of an overarching nature such that nature could exist without human beings, but not vice versa.

In general an acceptable form of anti-anthropocentricity for critical realism depends upon a three-fold set of relationships:

1 the ecological asymmetry, i.e. the existential independence of nature from human beings;
2 the essential dependence of human beings on nature; and
3 the natural character of human beings, i.e. the non-dualistic nature of the relationship between human beings and nature, such that human beings are natural, i.e. not only depend upon, but are constituted by, nature (and so are natural).

Together these three characteristics define a view of humanity as a species constellationally overreached and contained by (or englobed within) nature.

The theme of stratification within basic critical realism is itself a complex one. To the simple distinction between the domains of the real and the actual we need to add the sense of the multi-tiered stratification of nature, so that the distinction between the domains of the real and the actual is revealed by the development of science to be iteratively reapplicable. Then, in this context of the multi-tiered stratification of nature, we need to add a further sense of the *emergence* of higher order states and properties of things from lower order states and properties. Here it is crucial to distinguish the possibility of a diachronic explanatory reduction from the existence at any moment of time, i.e. synchronically, of emergent powers and states. There are three criteria for the existence, synchronically, of such emergent states and properties:

(i) the unilateral dependence of the higher order on the lower;
(ii) the taxonomic irreducibility of the higher order level (in describing and understanding higher order phenomena) from the lower order states; and, finally,
(iii) the causal irreducibility of higher order states in the generation of lower order phenomena.

We shall see in a moment how emergence illustrates a form of negentropy, which is also at work in the figure of progressive or dialectically sublatory change, which, albeit only partially and conditionally, qualifies the force of the entropy law of thermodynamics, introduced into ecological debates by Nicolas Georgescu-Roegen.

A fourth feature follows from evolution or the becoming, diachronically, of such synchronically emergent states and properties. This is:

(iv) the condition that the higher order state is implicit as a possibility in the lower order state, in which we could also say it was 'enfolded'.

Points 1–3 above, together with (i)–(iv) above, begin to define the form of an acceptable anti-anthropocentricity. This is sketched in *Scientific Realism and Human Emancipation* (written for the most part in the late 1970s though only published in 1986) as a *developing integrative pluralism*, in what is also characterized as a metaphysically Copernican–Darwinian world.

The distinction between open and closed systems (in CR the case of isolated systems is clearly differentiated from the case of constant ones[4]) is important for both CR and NEP. Indeed, as Karl Georg Høyer notes in Chapter 4 below, one could say that the misapplication of technologies or results, developed or tested under closed conditions, to the open systemic world has been a major source of ecological error and indeed disaster. What CR attempts to do, however, is show under what conditions the application of closed systemic results to the open systemic world is possible and valid and under what conditions it is not. It is valid when what we have is a statement about the non-actual real, i.e. when it specifies the operation of a mechanism or structure, a tendency irrespective of its outcome, at the level of the real; not when what we have is or depends on a description or prediction of the occurrence of an event or state-of-affairs, at the level of the actual.

The necessity for interdisciplinarity follows from the combination of complexity and emergence, i.e. from the ubiquity of open systems characterized by entities and structures with emergent powers, liabilities and tendencies. Interdisciplinarity is of course also very much a theme prevalent in the NEP literature. But the conceptual underdevelopment in the NEP reflection on the stratification of nature has tended to result, from a CR point of view, in insufficient recognition of the role of *disciplinarity in interdisciplinarity*. For disciplinarity is an essential ingredient in interdisciplinarity, at least in so far as the distinct structures which are operative in the production of the phenomena to be explained need first to be understood in their own terms, before they can (and in order to) be correctly integrated in an interdisciplinary account. Thus in the NEP tradition we have the theme of interdisciplinarity as practised by 'generalists', or even 'super-amateurs' (as distinct from specialists). I would prefer to say that it is dependent on the work of *cross-disciplinary understanding* and *effective epistemic integration*, mediated if necessary by the immanent critique and development of recalcitrant disciplines.

Now in general there are two distinct types of ontological grounds for interdisciplinarity: complexity and emergence, stressed at least in the first instance by critical realism, and holistic causality, emphasized by Nordic ecophilosophy. However these two modes can be shown, if rigorously analysed, to presuppose each other. Thus, on a CR account,[5] whereas for multidisciplinarity one requires the emergence of levels, for interdisciplinarity, in addition one requires the emergence of outcomes or results, i.e. a synthetic or novel outcome. This would seem to presuppose some degree of internal relationality between the interacting mechanisms. Similarly one only gets the need for multidisciplinarity or interdisciplinarity in the NEP argument, if the holistically interacting entities are indeed constituted at different ontological levels. So in practice complexity and emergence, on the one hand, and holistic causality, on the other, presuppose each other. CR has fully analysed both emergence and holistic causality[6] and in *Interdisciplinarity and Climate Change*[7] and elsewhere holistic causality assumes due importance in a second round of amplification of the basic RRREIC model of multi-mechanismic concrete explanation. In DCR of course emergence straddles 1M and 2E, while holistic causality belongs to 3L.[8]

Once we get an ontological picture of the interdisciplinary case, as in the notion of a 'laminated system',[9] there remains of course the question of how we secure this in epistemological practice. For CR this requires cross-disciplinary understanding plus effective epistemic integration,[10] buttressed (if necessary, e.g. to secure referential overlap) by the immanent critique and development of wayward disciplines; whereas NEP would tend to rely on gifted 'super amateurs' or 'generalists'. However a degree or specialization for the discovery and detailed description of the depth structures of the various sciences would seem to be inevitable, the important question being how to counter-balance it with sensitivity to the inputs of other disciplines. Here it is important to try to identify currently immanent tendencies that critique can latch on to guide/inform praxis. This is because, for an explanatory critique to be effective, it always needs to be accompanied both by a degree of concrete utopianism and a theory of transition. In this last, at least ideally, one should identify (whenever possible) emergent tendencies, which are already moving in the necessary direction.

In practice the transition to an adequate epistemic integration or an authentic interdisciplinary explanation can be enormously aided by creative conceptual work. Thus different levels of reality can begin to be coherently related to each other through the notions of four-planar and seven-scalar social being.[11]

Critical naturalism

The macro-dualisms which critical naturalism critiques and resolves, namely those of structure and agency, of methodological collectivism or holism versus individualism and of meaning and behaviour or law (and in a meta-theoretical key, of hermeneutics versus positivism, or anti-naturalism versus naturalism) are ultimately grounded for critical realism on the critique and resolution of certain micro-dualisms, namely between mind and body, reason and cause and fact and value, and also theory and practice. Crucial here is the notion of intentional causality, which links mind and matter, bypasses manichean dualism and alone potentially makes reason, and always, everywhere, actually makes thought (whether rational or not), efficacious in the world.

On the whole, while NEP does not theorize intentional causality, most of the work of Nordic ecophilosophers in the social domain is fully consistent with the emphases of the transformational model of social activity, oriented as it is against voluntarism and reification alike. In particular the Nordic ecophilosophers never doubted the need for social and collective action, as well as individual responses to the ecological crisis.

The theory of explanatory critique

Moreover, the Nordic ecophilosophers have on the whole accepted that facts can provide a rational basis for values, and they have engaged in forceful proto-explanatory critiques. However, the debate about values has left a problem, which turns on the fact that while human ethics is, in virtue of being action-guiding, necessarily concerned with the principles that govern human conduct, it cannot surely follow that these principles should encapsulate a false anthropic understanding of the place of human beings in the cosmos. In a re-enchanted world, value must exist in the world, whether human beings do or not. Value, like ontological truth, depends only on the possibility of language and ethics, a possibility which we know to be real, even if it were to be unactualized, or to become unactualized again, as it once was.

Section 2: dialectical critical realism

Emergence and dialectic; emancipation and shedding

The first law of thermodynamics says that energy can neither be created nor destroyed. This is commonly known as the *law of conservation of energy*. The second law of thermodynamics states that when energy is used it moves towards increasing disorder and chaos. This is commonly known as the *law of entropy*.

At first blush, the second law would seem to be contradicted by a number of phenomena in the biological and social spheres which appear to take on a negentropic form. In particular some changes and reorganizations appear to be moves in the direction of increasing order, inclusiveness, comprehensiveness or generality. These include:

(a) emergence and evolution in the biological sphere, which depend upon the energy of the sun or other free but finite natural sources;
(b) technological progress in the production, distribution or exchange of material goods or services;
(c) dialectical sublation or progress in social organization or ideas, and indeed learning generally.

Now so long as consciousness, thought or discourse is regarded as a free resource, then (c) does indeed assume a negentropic form. And ecology as an orientation or discipline, or learning or applying it, only makes sense if these activities are indeed negentropic. But in practice, consciousness or ideation is never an infinite resource, and it is always in the human world dependent upon finite material resources; as such the history of at least human consciousness and self-consciousness must always be seen under the aspect of, and as ultimately constrained by, the biological geo-history and circumstances of the human species. This is in turn inscribed within the planetary history of species and genera, itself embedded in the geo-physical history of the solar system, and ultimately the universe. However, that said, in so far as changes in ideas and social forms consume no more energy, they do indeed escape the consequences of the second law of thermodynamics.

However one cannot be so sanguine about alleged technical progress in the case of material, as distinct from purely conceptual, goods. Because what is technical progress or greater efficiency cannot be defined in economic terms without being defined in terms of its contribution to human well-being as a whole, and this cannot be defined from a non-anthropic standpoint except in terms of its contribution to the well-being of other species as well and the planet as a whole.

Indeed, as is now slowly becoming recognized:

(a) there is no question of maintaining current levels (and forms) of economic activity and current expectations about Pareto-optimal growth (whereby no one becomes worse off) within environmentally sustainable limits (not least for a rising human population as a whole). Rather
(b) what is required is net degrowth together with a radical redistribution of wealth, resources and energy – use for the planetary human population as a whole (plus stabilisation of its size at a tenable level). This will be taken up again towards the end of the chapter, when we shall see how the philosophy of metaReality entails a policy of shedding and radical simplification (de-complication in Sigmund Kvaløy's terms) for enhanced well-being-development for the well-off citizens of the well-off 'North'.

1M–7Z/A

In post-basic CR, there are altogether seven levels through which ontology is progressively deepened. The first level or moment, 1M, corresponds roughly to that of pre-dialectical, basic critical realism (at least *qua* transcendental realism).

- 1M thematizes being as such and being as non-identity, difference and structure.
- The second level or edge, 2E, thematizes being as containing absence and negativity, and thus as involving change, process and development.
- The third level, 3L, thematizes being as together or as a whole, and so being as a totality.
- The fourth level or dimension, 4D, treats being as incorporating transformative agency.
- The fifth level or aspect, 5A, presupposes being as involving reflexivity, and comprehends it as spiritual and involving inwardness.
- The sixth realm, 6R, understands being as re-enchanted, incorporating meaning and value.
- The seventh zone or awakening, 7Z/A, treats being as prioritizing identity over difference and unity over antagonism and split, and as involving non-duality.

In DCR we are only concerned with 1M–4D.

1M

1M takes its starting point from the critique of the Humean theory of causal laws (with its repetitive and shallow anthropism), together with the critique of the epistemic fallacy and an emphasis on the open-systemic nature of the world, entailing, given the phenomena of complexity and emergence, the necessity for interdisciplinarity. The critique of empiricism and neo-Kantianism can be developed to identify the effects of a complementary triangle formed by the classical paradigm of action (as atomistic and mechanical), the celestial closure apparently achieved by Newtonian mechanics and a model of (tacitly gendered, propertied and European) man, as egocentric and unrelated to other beings.[12] What lies behind the anthropocentricity so often detected by ecologically sensitive critics of orthodox or mainstream philosophy is in fact an *egocentricity*. This egocentric atomism together with an abstract universality lies at the heart of the philosophical discourse of modernity. We can see this discourse signalled already in Descartes' famous *cogito* 'I think, therefore I am'. In prioritizing thought over being, Descartes prioritizes epistemology over ontology and thought over body and emotion (not to mention the possibility of spirit). In prioritizing the 'I', Descartes privileges human beings over other species (and nature generally), and the speaker over society and other people.

I have described the development of the discourse of modernity in detail elsewhere[13] but a brief resume may be in order here. The philosophical discourse of modernity is structured by an atomistic egocentricity and an abstract universality. I have already noted how, in the Cartesian *cogito*, a self-defining subject is set against a given world, so that subject is opposed to object, and indeed other subjects. This quickly leads to the Hobbesian war of all against all, and a situation in which we are related to things outside ourselves only by Humean appetites and aversions, desire and fear. The defining couple of this discourse, atomistic egocentricity and abstract universality, quickly generate other philosophically significant and pernicious properties, including an incomplete totality and lack of reflexivity, unilinearity, followed by judgmentalism and disenchantment, (reductionist) materialism, formalism and functionalism, and above all ontological monovalence.

The philosophical discourse of modernity may be conveniently divided into five phases, each marked by a revolutionary upheaval or transformation. The classical discourse of modernity is marked by the English civil war of 1640–60 and the French revolution of 1789. The phase of high modernism may be associated with the revolutionary upheavals of 1848 and 1917. The theory of modernization may be associated with the momentous events of 1945, 1947 and 1949. Postmodernism, together with the rise of the new social movements, can in turn be related to the revolutionary upsurges of 1968 and the early 1970s. The fifth phase of the discourse of modernity, which I call bourgeois triumphalism, followed on quickly from the collapse of the Soviet bloc in 1989–91. This goes together with a resurgence of fundamentalism, both market and religious, and chauvinisms of various kinds.

I have further argued that this phase can be differentiated into three sub-phases, with the first sub-phase of globalization ending with 9/11, and the second sub-phase of the War on Terror ending with the credit crunch of 2007–8. We are now entering a period of multi-polarity, with new possibilities and challenges.[14]

2E

NEP, most comprehensively in the work of Torsten Hägerstrand, emphasizes the irreducibility of place and tense, the concrete country cousins of abstract space and time, most specifically in concepts such as that of landscape or (geo-historical) trajectory. Dialectical critical realism is very sympathetic to this, and develops its own concepts such as that of a rhythmic or the tri-unity of space, time and causality and the presence of the past (or future) to designate the irreducibility of organic tensed, as distinct from reversible clock, time and of concrete places as distinct from abstract spaces.

However, at 2E above all, dialectical critical realism engages the critique of ontological monovalence, and together with it, the endism and triumphalism which has been so characteristic of modernistic discourses. The critique of ontological monovalence is a major metacritical gain of DCR, presupposed in practice by our notions of real change and intentional agency, but necessary also (or so I have claimed) for any being at all.

Indeed I have argued that the absence of absence is one of the three great failings of Western philosophy, together with the epistemic fallacy and what I have called the 'Platonic-Aristotelian fault line', which is an index of actualism and the de-stratification consequent upon the epistemic fallacy.[15]

However, it seems to me now that there is another failing of Western philosophy, which is almost as important as these three – so much so that one could almost style it as the fourth great error, though it also follows pretty immediately from the unholy trinity. This mistake is consequent upon the *atomicity* typically invoked within the Western philosophical tradition, i.e. the absence of any space within the entities it posits or invokes. This atomicity, or *lack of any space within* or interior, results in the absence of any concept of inwardness, and a corresponding fetishization of the outer 'external' world. Atomicity here chimes in with egocentricity and results in an ideology of what we could call 'externalism', which includes and depends on 'extensionalism'. This fetishism of outwardness results in the elision or collapse of *inwardness*, including inner depth, reflexivity or even a coherent concept of the self. Externalism at its most extreme gives rise to behaviourism. Outlawing internal relations, it precludes hermeneutics (and empathy) and results in a shallow depthless view of human beings, and motivation, as involving the absence of an interior, of *any space within.* There is no room for intentionality, for the plane of the stratification of the embodied personality, for indeed any psychology worthy of the name. And the whole domain of metaReality is rendered invisible and ineffable. Indeed, it can only be revealed by a kind of 'symptomatic reading', which might be termed an *ecstatic reading*, in which the inner can be seen to stand out for, in and by itself and its kind.

3L

The first concept of the inward is however simply that of internal relations, which is the philosophically most important concept of 3L. Internal relations are the basis for holistic causality and the grouping of things together and as a whole, and therefore into totalities, and thus into distinctive and identifiable natural kinds, including the concrete singulars of the world. Together with this circle of concepts goes concrete universality and singularity, constellationality, reflexivity and the unity of theory and practice or what I have called 'seriousness'.[16]

4D

The critical realist theme of transformative praxis resonates with the NEP theme of meaningful work at 4D. And here there are interesting parallels with the DCR critique of irrealism, structured around the combination of anthroporealism and transcendent realism.

At the heart of anthroporealism lies egocentricity and abstract universality, resulting in a repetitive uniformity and inner emptiness, a Macdonaldized world. This vista of the incessant production of sameness and nothingness, however both always leaves traces of something determinate, a definite footprint and always (or so the philosophy of metaReality will claim) depends upon human ingenuity and teamwork, definite degrees of inter-human solidarity and skill.

The necessary complement to anthroporealism, the reduction of being to a superficialized knowledge, is an imaginary transcendent realism, required to compensate for it. At the heart of transcendent realism lies various *imaginaries*, in which reality TV, a lottery ticket, the cult of celebrity, window gazing in the shopping malls of the idle 'filthy rich', the promotion of one's football team in a different culture the other side of the world, co-mingle in a fantasy world of Disneyfication, the imaginary rewards for real drudgery.

However, just as the world of the repetition of the same leaves definite traces and residues on material reality, so does the surrogate world of fantasy necessitate and depend on definite psychic states and real changes, however seemingly shallow (at least from the outside), in real human beings.

The philosophy of metaReality goes further than NEP in arguing that within the world of exploitation and illusion, of demi-reality, of the industrial growth society, or of Sigmund Kvaløy's corrosive ACID, there is a level of the metaReal which is its basis.

Section 3: philosophy of metaReality (PMR)

Both Arne Næss and Sigmund Kvaløy have been concerned of course with the inward and spiritual in the development of their ecophilosophies. However, as Trond Gansmo Jakobsen has pointed out in his chapter in this book, their ecophilosophies depend on radically different cosmologies. The emphasis in Arne Næss' case is very much on considerations drawn from 3L and to an extent 7Z/A in the IM-7Z/A deepening of ontology, while Sigmund Kvaløy gives a very second edge (2E) analysis. It is of course a philosophical advantage of critical realism that it can accommodate both.

If basic CR allows the non-aporetic situation of change, it is the claim of DCR to be able to sustain a coherent account and analysis of change, namely as the absenting of something formerly present and/or the presenting of what was absent. In PMR our concern shifts to bringing it about. It is the being of the becoming, the change you would realize or bring about.

The philosophy of metaReality is not an otherworldly philosophy; and the spirituality it identifies is immanent, and indeed actual (not just possible).[17] However, the conceptions of identity and unity in PMR are very different from the conceptions of identity and unity prevalent in mainstream philosophy. For instance, the identity involved is rich, differentiated and holistically developing.

There are three senses in which social life presupposes the priority of identity. These involve respectively: identity as non-duality, which is the mode of constitution or reproduction and transformation of social life; identity as basis or ground state; and identity as deep interior. There are in turn four forms of non-duality: transcendental identification in consciousness, transcendental agency, transcendental holism or teamwork and the transcendental self. And the self for its part must be analysed in a tri-partite way. On this analysis, it consists of an absolute ground state, a relative and shifting embodied personality and an always illusory ego. This sets for human beings the twin goals of achieving or restoring consistency between the embodied personality and the ground state and the elimination of the illusory ego.

When the embodied personality of the self is inconsistent with its ground state, the intentionality of the agent will be split. Self-realization depends on the unification or restoration of consistency between embodied personality and ground state. Self-realized human beings may be said to be 'negatively complete', but positive completeness depends additionally on the abolition of all oppressive or master-slave-type relations and the elimination of all heteronomous states on all four planes of social being, together with the clearing up of their material residues. Together the criteria for negative and positive completeness deepen the criteria for the satisfaction of a 'eudaimonistic society', by elucidating further necessary conditions for human well-being and flourishing. This in turn allows us to identify the ways in which the project of universal self-realization both extends and is required by the axiology of freedom in DCR.

The foregoing is then developed in an argument in which the domain of the spiritual or metaReal is seen to be the ubiquitous presupposition of the practices of everyday life, which are conducted for the most part in terms of the categories of the world of duality and under the influence of that exploitative, oppressive and categorically false part of it which I have called 'demi-reality'. This practical presupposition may be illustrated by the way in which trust underpins commercial transactions, or the way in which war presupposes some peaceful activities, a degree of peace, but not vice versa, and the way in which we use our creative ingenuity to stay stuck or addicted, or our loving concern to fuel a host of negative emotions, from jealousy to hatred. All these forms exemplify an *axiological asymmetry* in which we can have the normally under-recognized good or 'heavenly' state without the eye-catching and dominant bad or 'hellish' state, but not the other way round.

This has profound implications for our understanding, critique and replacement of the demi-real world of alienation, oppression and growing inequity and injustice.

Thus though the resonances are strong between NEP and PMR, especially perhaps in the case of what Arne Næss called 'ecophilosophy T', what needs to be unequivocally registered is that social reality could not exist for a moment without its spiritual infrastructure, which is its basis, mode of constitution and deep interior, or so PMR argues. Thus there is a limit to the production of nothing, non-places, non-things. It is the labour, creativity of real people as it involves real places, landscapes, trajectories, sinks and sources; a labour which involves real compassion, solidarity and trust. The theorem of the impossibility of ecological neutrality must be accompanied by that of the *inexorability of human engagement*. Nothing truly can happen, at least in the social world, without it.

Thus there is no society without human agency, and no human agency without both non-duality and the ground state.

Together with the production of nothing (in NEP critique) in a Macdonaldized world goes the consumption of nothing in the Disneyfied world of fantasy. But to work, the fantasy must touch a human heart, evoke a cathected aeffect, however shallowly. The human being must operate the machine and choose to play out one or other kind of narrative as the story of her life.

If the metaReal is an unacknowledged but real substructure of social life, for the most part conducted and conveyed at the level of, and in the idiom of, the world of duality but under the sway/dominance of the demi-real, then we can say all of three things. We have the *metaReal* (a) *in* (b) *under and* (c) *against the actual.*[18] The metaReal is certainly in, and a part of, though a largely unrecognized part of, the actual; it is certainly dominated by the world of duality, and the world of demi-reality within the latter; but, especially if developed in the concrete utopian imagination, it depicts and holds out a vision, albeit one only instantiated in microcosm, of a better way of doing things, of an alternative order of human being and social life.

There is another consequence of great moment for our crisis today. PMR affirms that the only way of avoiding a split intentionality, and the failure and unhappiness that inevitability ensues, is to eliminate the inconsistencies between the ground state, which one cannot lose, and the rest of one personality. This will involve the shedding both of the ego, the illusory sense of a separate 'I', and of the heteronomous characteristics in the embodied personality, i.e. those features of it which are inconsistent with its ground state. This will inexorably involve a simpler, but inwardly richer and deeper existence.

This in turn will inevitably involve:

1 degrowth; and
2 the enhanced use of underdeveloped powers of humanity, together with aids to them such as IT technology.

1) Karl Georg Høyer and Petter Næss have elegantly argued for the importance of uncoupling economic growth from the objective of environmental sustainability, suggesting that trying to marry these objectives presupposes 'an unrealistically high degree of eco-technological efficiency increase' (Chapter 14, p. 218). The upshot is a double impossibility theorem: it is not possible to have growth and ecological viability, and because it is not possible to have capitalism without growth, it is also not possible to have ecological viability with capitalism.

An end to growth is not however the same as an end to development. In two senses – first on the terrain of economics, in the context of a world of growing inequalities, there needs to be a radical redistribution in favour of the poor, sick or disabled, the young and the old, the oppressed and the needy; together with this redistribution we must encourage a flowering of new transitional and ecologically friendly technologies, discovered or invented and produced and functioning without any overall increase, and preferably with a decrease (the larger the better), in the use of energy; and with this also must come a profound re-organization of our life-long learning, so that we learn to enjoy and share not just information, but knowledge, ideas and wisdom, and to become adept in the skills of the imagination and the arts of empathy. Thus in learning to become one with the other one is (or thinks one is) not, we learn that we need no longer bear the pain of the repression or exclusion of a part of oneself. Secondly, development will take place at the level of the greater fulfilment of our individual and collective human nature, consistently with our responsibilities to other species and unborn generations, that is, towards the goal of universal human flourishing, and of the free flourishing of each as a condition of all.

2) This can be illustrated by considering the phenomenon of hypermobility, incessant travel in fuel-guzzling cars and planes. Thus is there any real need for most, perhaps 95 per cent, of the passenger flights from say Northern Europe to the continent of Australasia? If one wants to go to Australia, perhaps to take in the Sydney Opera House or Gold Coast in Queensland or Uluru, why not watch a film or video of them, or go on the Internet or make a telephone call to a friend there or read a book or talk to and empathize with someone who has been there. Of course one will have to do a bit of inner work, delve into one's inward space a bit, to use one's imagination. But how much more rewarding, and better for oneself and the planet, including of course Australia, than a 20-hour flight! Again, who really needs to go to the Grand Canyon, at least just to *see* it, when they can experience it on film? Our aided imagination, and consciousness generally can take the place of an enormous amount of our current travel.

Two final questions remain:
First, are there any intrinsic limits to alienation in four planar social being? The philosophy of metaReality identifies a limit to the alienation and crisis we experience at each of the four planes of social being. At the plane of social interactions between people, this comes from the irreducibility of transcendental identication in consciousness in social interaction, manifest in the consideration that even when we do not agree, we must to some extent understand each other (even to mark our disagreement); and even where we do not understand each other, we can at least make out the words the other is uttering; and even when we cannot do this, we can at least hear them. Similarly the person who is split ten different ways or hears ten different voices, at least *hears* the different voices – there may be ten different voices, but there is only one listener, there is one person who is split, a real self on which to build, from which to bootstrap. This then is the limit at the plane of the stratification of the embodied personality. The limit at the plane of material transactions with nature comes from the fact that human beings are natural beings. Nature is not apart from us, we are a part of it. The destruction of nature is not only murder, but suicide, and must be treated as such.

The case of the plane of social structure *sui generis* is more complex, and best discussed by reference to an example, such as the recent credit crunch. The solution here is the re-embedding of money in the real economy, and the re-embedding of the real economy in the social and political infrastructure on which it depends, together with the re-embedding of the social infrastructure in its metaReal spiritual substructure. In the case of the social structure and our transactions with nature, more than individual intentionality and agency is required. We need collective decision-making and action of course, and here PMR invokes the the axioms of universal solidarity and axial rationality, which serve to bridge alleged incommensurabilities and opposed interests.[19]

Second, is there a sense in which, as Trond Gansmo Jakobsen charges, NEP is tendentially reductionist, in so far as it is committed to a basically mono-disciplinary – and ecological – account of reality, albeit it is accepted that an ecological account will be a holist one? Here I leave the reader to peruse his argument in Chapter 13. As I understand it, his argument is that critical realism, at least in its full 1M–7Z/A development, offers a more comprehensive philosophical ontology and a fuller range of options than the ecological paradigms which at least Arne Næss and Sigmund Kvaløy relied on; and that moreover at the level of substantive scientific ontology, their ecological vision needs to be supplemented by the ontologies of other sciences, disciplines and practices, including in principle folk wisdom and popular imaginaries. This will be true especially, but not only, in the case of those planes of social being other than that of our material transactions with nature.

Certainly no one would deny that NEP has focused on ecological aspects and issues. Clearly these are not the only issues and aspects, even if, in so far as they represent an absolute constraint for humanity, they can be said to be conjuncturally the most important ones, the *sine qua non* for any other good idea or dream. However, that said, what we have to consider is the possibility that capturing and describing the non-ecological aspects will unlock the mechanisms, at the level of human agency and social structure, to a transformed transformative practice in relation to our natural environment, based on principles of care and responsibility for our natural environment, of which we are currently such a destructive part.

Notes

1 MinGyu Seo, 'Bhaskar's Philosophy as Anti-Anthropism: A Comparative Study of Eastern and Western Thought', *Journal of Critical Realism* 7(1) 2008, pp. 5–29.

2 See *Scientific Realism and Human Emancipation*, 1986/2009, Routledge, London, p. 130; *Dialectic: The Pulse of Freedom*, 1993/2008, Routledge, London, p. 160.

3 See *Scientific Realism and Human Emancipation*, p. 141.

4 See *A Realist Theory of Science*, 1975/2008, Routledge, London, p. 76.

5 See my 'Contexts of Interdisciplinarity' in *Interdisciplinarity and Climate Change,* edited by Roy Bhaskar, Cheryl Frank, Karl Georg Høyer, Petter Næss and Jenneth Parker, Routledge, London, 2010, Chapter 1.

6 See e.g. *Scientific Realism and Human Emancipation* and *Dialectic: The Pulse of Freedom.*

7 See Chapter 1.

8 See *Dialectic: The Pulse of Freedom*, Chapters 2.7 and 3.8.

9 See Roy Bhaskar and Berth Danermark, 'Metatheory, Interdisciplinarity and Disability Research: A Critical Realist Perspective', *Scandinavian Journal of Disability Research* 8(4) 2006, p. 288.

10 These are meta-theoretically underpinned by the principles of universal solidarity and axial rationality but in practice presuppose major changes in our educational system, in terms of much later and less than full specialisation, and the encouragement of hermeneutic sensitivity and the empathic arts generally.

11 See Bhaskar and Danermark, 'Metatheory', p. 289; and 'Contexts of Interdisciplinarity', pp. 8–10.
12 See *A Realist Theory of Science*, p. 198.
13 See *Reflections on metaReality*, Routledge, London, 2011, Chapters 1 and 4.
14 See Roy Bhaskar with Mervyn Hartwig, *The Formation of Critical Realism*, Routledge, London, 2010, Chapter 9.
15 See *Dialectic: The Pulse of Freedom*, Chapter 4, *Plato etc.*, 1994/2009, Routledge, London, Chapters 8–10 and Appendix, and cf. also Alan Norrie, *Dialectic and Difference: Dialectical Critical Realism and the Grounds for Justice*, Routledge, London, 2010, Chapters 6–8.
16 NEP has always encouraged a close relationship between theory and practice.
17 See *Reflections on metaReality*, 2002/2011, Routledge, London, *From Science to Emancipation*, 2002/2011, Routledge, London and the *Philosophy of metaReality: Creativity, Love and Freedom*, 2002/2011, Routledge, London; and also my forthcoming *A Brief Introduction to metaReality,* Routledge, London, 2012. See also Mervyn Hartwig and Jamie Morgan, *Critical Realism and Spirituality*, Routledge, London, 2011.
18 To borrow Alan Norrie's lapidary formulation in his *Dialectic and Difference: Dialectical Critical Realism and the Grounds for Justice*, Routledge, London, 2010, see Chapters 5–7.
19 See my 'Theorising Ontology' in *Contributions to Social Ontology*, ed. Clive Lawson *et al.*, Routledge, London, 2007, pp. 200–3, 'Contexts of Interdisciplinarity', in *Interdisciplinarity and Climate Change*, ed. Roy Bhaskar *et al.*, pp 18–19 and *The Formation of Critical Realism*, p. 198.

3

TECHNOLOGY, NATURE AND ETHICS

Arne Johan Vetlesen

I

In his book *Ecology, Community and Lifestyle*, Arne Næss (1989: 103) writes:

> When a technique is replaced by another which requires more attention, education, and is otherwise more self-engaging and detached, the contact with the medium or milieu in which the technique acts is diminished. To the extent that this medium is *nature*, the engagement in nature is reduced in favour of engagement in the technology. The degree of inattentiveness or apathy increases and thus our awareness of the changes in nature caused by the technique decreases.

Næss' observation makes me think of the machines nowadays used in industrial logging. I am reminded of the time I came across a vacant vehicle of the kind used in Norwegian woods today, as well as in the Congo and in the Amazon. I climbed aboard, made myself comfortable in the driver's seat (it was a cold autumn day and raining), and guess what commanded my attention?

The answer is the computer placed above the steering wheel. Looking at the screens and pressing the appropriate keys (that is, doing exactly what I do right now, sitting in my home office on a winter's day), I would be able to direct the grab toward the tree selected for felling, to make sure that all the branches be removed, to cut the tree, and to place it onto the pile of timber on the plane. I didn't actually carry out these operations, of course. But I have witnessed how it is done: it is showcase of modern, efficient technology. What a generation ago would have taken hours to accomplish, is now effected in a matter of minutes. The selected tree may have taken hundreds of years to grow, it may be forty meters tall and weigh hundreds of kilos, thus requiring considerable physical effort and experience-based skill to be taken down, by saw or by axe.

'Self-engaging' and 'detached': Næss' words ring true. Felling large trees when sitting comfortably inside a latter-day forestry vehicle makes for what I term *abstraction*. My field of attention shrinks to keeping a keen eye on the computer in front of me, combined with the occasional look outside to check that the branches are indeed removed, the tree indeed cut, and so forth; in a word, to check that the orders I feed into the computer effect the desired operations upon the outside world. The task at hand has lost – or overcome – the physical effort its distinct elements once called for: selecting the tree, deciding in what position to enter the saw, considering the optimal direction for it to fall without mingling with other trees, then going about the actual sawing, changing position, taking a break, completing the felling, making sure the tree falls in the desired direction. Doing this in the low-technology manner means (meant) for the carpenter to engage in direct contact with the tree at hand; a contact of physical-bodily-sensuous nature. Each tree would be different: different position, different height, different shape. Each would call for judgment concerning where, when and how. The felling of one tree would prove faster and less laborious than expected; the felling of another the opposite. The sheer physical effort required to complete the job, the time it would take, combined with the absence of perfect predictability and calculability of each operation, would make each tree a *particular* challenge and a particular experience. No two trees are exactly alike. They only become alike once felled, forming a pile of equally tall and equally heavy logs to be taken away either by horse and sledge or – closer in time – by tractor.

This makes for an illuminating contrast with the high-tech machinery used in modern industrial forestry. Perhaps *self-absorption* is an even more apt description than self-engaging and detached (Næss). They all point in the same direction, though. The principal difference involved in my example (which I propose is highly representative of the kind of shift under scrutiny) is from one sort of relationship to an entirely different one. The parties to the relationship change: from one between the individual carpenter and the particular tree he is engaging with, saw and axe acting as direct physical extensions of his bodily movements, to one between the worker and his computer screen. Whereas all the feedback in the former case would issue from the tree – the tree now seen, touched, heard and smelled; the tree yielding or resisting, the tree still standing or, at long last, falling – the locus of feedback to which the modern worker's direction is directed is the computer. The man–tree relationship has been substituted by a man–machine relationship. So it is not a matter of man's relating-to the tree being mediated by a machine (computer) that literally is an in-between between subject and object. Rather, and more profoundly, it is a matter of the subject–object (outer reality) relationship being *replaced* by a man–machine (technology) one. The worker works on the machine which, according to the instructions feeded, works on the tree. The machine – computer directing the saw-machinery – works physically on the tree, the worker does not: he does not touch it, does not smell it, he hardly hears it (falling), he may see it (but he need not). The machine, in the given comprehensive and multi-functional sense, stands between the man and the tree, the subject and his object, making a direct (physical-bodily-sensuous) exchange between them all but redundant. The tree felled is not a concrete tree with a unique set of properties (location, age, height, weight, shape, look) with which the worker interacts. It has instead been transformed into an abstract entity, exchangeable not unique, operated upon via a technical device rather than sensuously and physically experienced.

The larger picture is this. Today nature no longer constitutes man's dominant reality. Whereas in the preindustrial era nature was ever-present and all-powerful and instigated a feeling of inferiority and fear on the part of man, nature has now lost that quality. Say 'wilderness' and the spontaneous association is something precarious, shrinking and threatened by extinction. Why this change? Because of the expansion of man's activities into nature, taking control by way of technology over what was previously beyond reach. Man-made infrastructure is turning what used to be unknown and untouched territory – wilderness – into so many industrially extracted and commercially exploited resources. At ever-increasing pace, we have confined what used to be nature into small parcels. Instead of fearing being overtaken by nature, technology in our era has enabled us to seize control over nature to an extent unimaginable to our forefathers. Hence Yupik Eskimos (and other indigenous people) with incredibility and apprehension refer to us Westerners as 'the people who change nature' (McKibben 1990: 84). Growing ability to control, to exploit for human purposes, has made us fearful of nature in a novel way: fearful not of nature as such and its intrinsic powers, but instead fearful of what we, thanks to our technologically enhanced powers of transformation have done to nature, of how it will backfire.

What counts as reality today is to large extent made up of artefacts whose very existence would be impossible without the workings of modern technology. What we use while carrying out our everyday activities is provided by technology combined with mass production: the computer I am now writing on, the cup I am drinking my imported coffee from, my table and chair, my mobile phone. I look around, and what I see carries so many signs of the man-made activity that went into making it. It is as if modern man increasingly encounters only himself when engaging with the outside world, meaning the products of his own technologically mediated activities. By domesticating the earth, we've domesticated everything that lives on it, creating a situation where nature is no longer nature and where 'we can no longer imagine that we are part of something larger than ourselves', to quote Bill McKibben (1990: 90). If there is still an outside, something other and still untouched by man, its continued existence as such will be at the mercy of our decision to allow it to still exist, be it as an island in an ocean of domesticated nature, which precisely is not nature as it used to be and what it used to mean. Species that used to be plentiful and thriving in their habitat, thus taken for granted, have in the course of a few centuries or a few decades become rare, therefore requiring a reversal of business as usual in order to survive qua species (Barnosky 2009). Dramatically, this also goes for the quality of the basic elements of nature: water and air. Both are now polluted in many regions of the world, especially in urban areas. Their quality as clean used to be taken for granted, indeed it constituted their essence: air meaning apt for breathing, water meaning apt for drinking. Due to human activities, the intrinsic nature of the basic elements of nature is being degraded. Thus it happens that clean water and clean air become luxury goods, commercial assets invoked in advertising for resorts for the well-off whose level of consumption has helped cause their degradation in the first place.

Reality as experienced and reckoned with today, then, means domesticated reality. What counts as real is identical with what is already formed, indeed tailormade for our use of it, according to our perceived needs as human beings. Experience of nature as it is prior to these processes has turned from the rule to the exception; it must now be sought out, and again there are signs that experiencing nature in its untouched qualities is becoming more privileged the rarer such qualities de facto become.

As a consequence of this development, shifts – rhythms – that are intrinsic to nature are obliterated for man thanks to technology. Day and night, light and dark, warm and cold, in a word: the transitoriness of the day as well as the seasons, have all but lost their once commanding impact on the activities of human societies. Today an increasing number of activities will take place at different times of the day; work and shopping need not be restricted to daytime; the light and the temperature in the offices or malls one visits are the same regardless of the time of day, the outside temperature and the season of year. The activities constituting our present way of life have become quasi-autonomous and as such seemingly independent and separate from the intrinsic structure of nature. Technology's thrust is to effect a wholesale obliteration of time and space, meaning the differences of and in time and space that historically helped define – facilitate and restrict – work activities in particular (harvesting being a prime example). The specific *when* and *where* of my location cease to make the difference it used to make to all generations preceding us. The trend made possible – made into our reality – today is for me to be able to pursue whatever activity I wish at any time and at any place. The upshot of modern technology, especially computer technology, in this regard is to effect a gigantic *rendering accessible* of everything, everywhere and at every time; and effecting this, per trend, for *everybody*, not only due to increasingly more objects being drawn into technology, but also due to increasingly more people (more cultures) relating to objects in a technologically mediated manner, whereby 'objects' denote both other humans (i.e. Facebook and other computer-generated social media) and the basic elements of nature.

Let us dwell on the latter. To relate to cold and warmth in a preindustrial way is to experience first-hand that the night is dark and cold, the day light and warm. Consider next the experiential quality of a piece of comparatively primitive technology such as the oven. I feed the oven as the last thing I do before going to bed at night. During the night it will burn out. Morning I experience as cold, though if not for the oven colder still. I need to start the oven anew, relishing the transition from feeling cold to feeling warm. Waiting for the oven to heat up is also to wait for the kettle of water I placed on it to start boiling, allowing me to wash and to make coffee. To be sure, in this low-tech setting technology helps to reduce the impact of some of nature's intrinsic transitions: it is cold in the morning, but not as cold as it would be without my oven. Nature's *given*, as it were, properties in the form of rhythmic shifts – from dark to light, cold to warm – provides the objective context for my activities; they make it meaningful – not merely necessary – for me to keep the fire going, etc. In short, I experience myself as a human being *relating to* and *interacting within* a certain setting, a certain rhythm of shifts from one state to another, the shifts in question being not of my making but something real out there confronting me, limiting certain activities of mine, rendering it necessary to do this and that in order to compensate for the full and as it were unmitigated effects I would otherwise be at the mercy of (the sheer coldness of the night absent a house and an oven).

Contrast this with living in a modern, technologically up-to-date house where the wood-using oven has been replaced by a thermostat-driven one. This panel oven is a self-regulating device. Its function is to keep the room temperature at a desired constant warmth. When I no longer experience whether it is dark outside (indoors it's light) and whether it is cold outside (indoors it's warm), what I have referred to as the intrinsic rhythmic shifts of nature have become insignificant, indeed unexperienced, by me. There is nothing in my surroundings – in what I sense and feel – to remind me of the shift from – meaning difference between – properties of nature such as day/night, light/dark, warm/cold.

So what? What does this contrast amount to, and what is its importance?

I said that technology annihilates the particularities of time and space. It is easy to see why this has made for what is deemed progress: a materially secured and more comfortable life for increasing numbers of people. To have helped accomplish this is no minor feat, and I do not intend to belittle its historical importance. However, the gains of this development being advertised everywhere, my concern here is with the kinds of loss that accompany the celebrated gains.

Technology allows for our liberation from the constraints of *distance* and the frustrations of *waiting*. It makes possible, and helps reinforce the craving for, instant gratification: not of this or that desire, but all of them. The brought about annihilation of time and space as ontologically given contrasts, as marking differences that impose limits and so *differentiate the possible from the impossible*, the doable from the undoable (not now, not here), carries with it a promise as well as a pretence of instant accessibility of everything in the world that might be desired by a human subject. This is what philosopher Günther Anders had in mind when, in his major work *Die Antiquiertheit des Menschen*, he explored how technology in general, and audiovisual mass media in particular, create a historically unprecedented form of *simultaneity*: the technologies we have created see to it that we exist in the world in such a way that we have the world come to us instead of us coming to the world. The medial simultaneity of all events – everything that happens can be made present to me here and now; everyone, regardless of where they are and who they are, can get to see the same, in the same now – is at the heart of the world we have developed: a universe where we no longer are at some particular place and at some particular time in a sense that would make a difference for the accessability of the event we wish to experience or the commodity we wish to buy (Meyrowitz 1985).

Technology's ongoing and never-pausing annihilation of time and space as significant and limit-marking points of orientation in our existence – as organizing principles for social organization and communal life, as for basic subjective states such as expectation and hope, and as markers for the doable versus the undoable – makes for emptiness. My room shrinks – is not expanded – into a global village in keeping with the speed with which technology-driven production, distribution and consumption overcome all placebound barriers, and the time horizon shrinks to a point where the now and the actually present amount to everything that is, and where everything that exists is forced to appear as contemporary and present in order to prove its existence, its worth. The psychic counterpart to this is the experience of an overwhelming *compression* of my spatial and temporal existence (Rosa 2005). The mode of experience in the world, of the world, that this encourages, especially thanks to audiovisual media and computers (themselves now carryable everywhere), is that *everything* to be experienced in this world in principle can partake of, can inhabit, my here and now. In a world, the self–world relationship prompted is that of the world at my feet: reality as the extension of my desires here-and-now.

This gigantic production (suggestion) of presence creates its own kind of absence and hence of loss. Escaping from view is every trace of origin and prehistory: say, where the commodity I consume comes from, how and by whom it is made and its trajectory from those particularities to those making up my circumstances as someone now consuming it. The sources and manifestations of the singularity of the particular thing (resource-cum-commodity; labour-cum-product) that I access (buy, eat, see, use) here and now are all but completely eclipsed. *What you see is what you get; what you get is what you see.* The everywhere-omnipresent world, the world as coinciding with my here and now in each instant, is the world as surface; a world devoid of a past; a gigantic compression of everything ascribed significance and worth to my actual and self-absorbed here and now. To be sure, the world has become bigger; to a degree unthinkable to my predecessors it has become accessible to me: what until recently was out of reach, mainly because physically distant, has come to be included in my possessions. Yet the world has become bigger only by becoming smaller: expansion in time and space, the bringing down of all hitherto encountered (and of necessity respected) barriers in both directions, has the shrinking of the world and the leveling of experience as its unintended effect. Having the world at my feet in this sense makes for the most primitive narcissism. In a situation where the aptness for partaking of my here and now is the quality needed to count as really real, as mattering, as making a difference to me, most of what is truly out there is doomed to fade out of view, hence of significance, hence of love and protection.

Let us see more clearly what this amounts to.

II

To appreciate what is at stake here, consider spending time in a part of nature where wilderness can be experienced. Here time and space make themselves felt. Sunrise marks a time for openness, for beginning; the rising sun illuminates the shapes and elements of the landscape and gives content to the distinct directions of the compass. The movement made by the sun in the course of the day marks the temporary sequence from the start of the day over the afternoon, evening and into the night. The sun provides a place for our activities: rising when it gets light, hiking in the middle of the day, finding an apt place to camp (access to water and firewood, etc.) before it gets dark in the evening, sleeping while it's dark. Nature is not only eminently temporal; it is also eminently spatial. The landscape – never landscape as such but always *this* landscape – has its peculiar shapes and limits, vegetation and animal life varying from one location to the next, as marked for example by altitude, thereby marking out so many ecological niches: some conducive to the life and flourishing of some species, other to that of others.

Whereas technology insists on rearranging the landscape so as to impose on it form dictated by human interests, disclosing what a given piece of territory contains in terms of resources fit for exploitation, a pristine landscape changes only according to its own dynamics and in keeping with the rhythm of the day and the seasons. Areas of wilderness are well-articulated on behalf of themselves and attest to their intrinsic dynamics. Nature doesn't need humans; as a biological, geological and ecological system it is self-sufficient. In its pristine state, nature is the counterpart to technology. To capture this quality of nature, the American philosopher Albert Borgmann invokes the Greek word *deiknynai*, which means to show, to point out, to bring to light, to set before one, and then also to explain and to teach. Deictic discourse is about something that addresses us in its own right and constitutes a centre by which we can orient ourselves. In this perspective, wilderness exemplifies a focal concern and is therefore a crucial counterforce to technology. The commodities that are the goal of the technological enterprise, facilitating their consumption over ever greater distances and at ever greater speed, invite us to use up an isolated entity (this chocolate bar) without preparation, resonance and consequence; as a buyer and consumer of commodities, I only relate to the here and now of using (enjoying, devouring) it; I do not engage with the entity's past or future. Technology's way of making something accessible and present, meaning at hand and to-be-consumed and thereafter forgotten, is by breaking the entity loose from its pre-existing and particular context: by abstraction. Technology works by fragmentation and isolation; it is anti-holistic, disruptive of and destructive to the dynamics of connection, relatedness and interdependency inherent to all life in nature. Moreover, and as observed at the outset, technology precipitates disengagement from bodily and sensuous interaction with outer reality and hence from the performance of skills as embodied. What we consume thanks to technology will fail to centre and illuminate our lives. The result is distraction, the scattering of our attention and the atrophy of our capacities. What technology has liberated us from, such as the toil and monotony of manual labour, entails a diminishing of the fullness of our repertoire as human subjects in the world (recall Næss' opening statement): there is a close and mutually reinforced correspondence between the shrinking of the outside world (or nature as a whole) into so many items for consumption here and now, and the shrinking of my repertoire qua subject. So it is not only that technology discloses to us a world less rich in differentiation, one reducing things to their surface and robbing them of depth by presenting them as so many discrete entities without a history and a context, breaking them loose from their original relationships and measuring them by reference to one (simplifying, unifying, commensurability- and exchangeability-imposing) standard, mostly that of monetary value, securing their conversion into commodities. The same sort of leveling also goes for the subject confronting objects in this manner, this being a two-way affair.

To repeat, technology's enormous potential for offering us relief (disburdening) from the care of things is commonly seen as liberation from toil. That this much-celebrated progress is bought at a heavy price is what deictic discourse in the sense intended is meant to help us recognize. Deictic discourse allows us to be guided by focal things, matters of ultimate concern that are other and greater than ourselves. Such a discourse cannot, and does not aspire to, control its subject matter. The elucidation of the features that belong to this discourse should be understood as the anticipation and recollection of real or concrete things that are their own warrant. Such discourse embodies a certain kind of attitude, and it has a certain force. What it helps disclose to us as existing out there, being what it is independently of our interests and projects, being sovereign and autonomous and full-fledged in its own right (a tree, for instance). Such an object, encountered in its own place, taken fully in by us in its own being, has the power to prompt in us feelings of enthusiasm, sympathy and tolerance. We come to feel grateful that there is such a thing. What we see is not entirely up to us, let alone of our making; it is precisely not a product of our intentionality, nor of our desire. Instead it displays the character of a *gift*: I feel lucky to have spotted the two hawks riding the sky above me. Their presence – more profoundly, everything they are, everything they do – is beyond my powers and my control. As opposed to initiating something, I am in a position of reception, of being attentive to, of seeking to be up to what it is I am seeing, hearing, touching, smelling, in short, to the fullness of particularities constituting the world as perceived from this place, right now. In short, it is a case of attending to the world outside me, as distinct from having selected items of the world coming to me thanks to technology.

Deictic explanation, explains Borgmann (1994: 181), elicits active assent in us to what it helps disclose to us; it moves us to act, teaching us what to do by telling us what is, and thus disobeying the is/ought and fact/value dichotomies – more deeply: ontological dualisms – handed down to us by the tradition of Western philosophy at least since Descartes. Deictic discourse rests its case not by subjective standards *à la* Hume's famous introspective summoning of our feelings of affirmation or disgust but by pointing *away* from the subject to the thing in question.

By insisting that the proper point of reference is the thing itself – that tree over there, that hawk up there; that particular place in that region – practitioners of deictic discourse go against the grain of the two types of discourse favoured in contemporary technological society: the subjectivist as well as the objectivist one, whereby the former is thought to be apt for matters of taste and morals (aesthetics and ethics, if you will), the latter for matters of scientific truth and technological usefulness. Thus, to my enthusiastic exclamation 'Wow, what a beautiful tree, what a majestic bird!', accompanied by gratitude that such things (still) exist and thrive in their original habitat, the response can always be '*You* think it's great, but I don't'. So is it really? Who is to decide in case of disagreement?

At a person-to-person level, the best I can do is to counter objections by encouraging the sceptic to have another look, presumably a *better* look, one succeeding in appreciating the qualities in the thing out there – tree, bird – as they address us as human subjects, engaging us in our fully-fledged capacities qua subjects. The very chance to experience something in its own right, and correspondingly the chances that we as the subjects so addressed will be willing and able to do that 'something in its own right' epistemic justice – to pay it the respect, admiration and protection due to it – are both critically endangered by technology's having become the new orthodoxy, the dominant character of reality. Part and parcel of this orthodoxy is the discursive hegemony enjoyed by the language and standards of validity, relevance and truth peculiar to economy, the flip-side to technology and its ally in the only type of discourse that nowadays is considered serious and objective by the powers that be. As has been commented on many times, conservationists – or Manwoodians – of today

> are forced to speak the language of those whom they oppose. This is precisely the language of usefulness. In their efforts to preserve the forest sanctuaries, they must remind science as well as governments that one day the abundant diversity of plant species that exist nowhere else but in the forests will prove useful and beneficial for such things as treating cancer or other diseases. They must contrive a thousand convincing or nonconvincing arguments in favor of the utility of forest conservation. For the moment this is the only language that has a right to speak, for it speaks of the 'rights', that is to say economic interests, of humanity. It remains to be seen whether one day a less compromised, less ironic language will become possible – a language of other rights and other interests, a language, in short, of other worlds.
>
> *(Harrison 1992: 124)*

Given this orthodoxy, the chances that deictic discourse will prevail seem meagre. The technological universe – the kind of reality we respect and relate to thanks to technology-driven disclosure of it and access to it as what simply is, what constitutes reality per se – is deeply hostile to speaking of something out there, in nature, as being something in its own right, as commanding respect and as having value beyond human-projected usefulness. So stubborn and deaf to other perspectives – languages and the whole worlds that, reciprocally and equiprimordially, go with them – is this hostility that attempts to meet it halfway are the rule not the exception. But as Borgmann (1994: 187) warns, defending endangered species or whole habitats in the name of formal characteristics such as complexity or diversity entails dangers of its own, as does the reference to stability and integrity found in the formal concept of homoestasis. His point is well taken: the dominant general and abstract characterizations of nature cannot be countered by accommodating the concepts of, say, ecology (the discipline) into the only discourse now considered to be taken seriously, i.e. that informing and informed by the technological conception of reality, validity and value. Instead of moving upwards, by way of formalization and abstraction, we should move downwards, meaning closer to the immediate and grounding experience of wilderness. Indeed, there is no substitute for this: at its most profound and basic level, type of discourse is about mode of being in the world (Heidegger). What is truly crucial is to avoid, or help abandon and relinquish, any distance obtaining between one's characterization and one's experience of nature.

To appreciate the gravity of the problem, we need to realize that it is not merely that technological discourse prevents us from recognizing what human-independent entities are in their own right, in their otherness, in their depth and splendor. It is also that the stronger and longer this discourse prevails as dominant and ubiquitous, the fewer will there be of the entities in question, since the practice driven by this discourse causes them to be eradicated from the surface of the earth. In short, the kinds of entities to which the deictic discourse corresponds, i.e. the entities constituting its proper universe, are as such threatened by physical extinction. Whereas on the one hand there is as a need to make people pause and to say 'Look at what's there, look at the wonders of wilderness and all the animals and plants and other species that go to comprise it', so as to prevent a fully-fledged subjectivity of attention and reception from lying dormant, there is on the other hand a situation of precariousness and scarcity when it comes to the availability of the entities in question. The more distanced to them and their proper ways of being – growing, thriving – we become, the more easily and unreflectively we become complicit in their destruction on a daily basis, shaped by the technology-driven being in the world and its peculiar (though avowedly detached, impartial and objective) meaning selective mode of disclosing 'reality'. And the further we have, historically and collectively, by accumulated effect, proceeded down this particular avenue, the fewer and rarer will be our chances to have this mindset of ours *corrected* by its alternative, namely a deictic discourse. We are running out of time. The need for deictic discourse to assert itself, to challenge the dominant discourse-and-practice, is growing more urgent each decade, year, month. Yet the objects it relies on to enable a first-hand experience of such an alternative mode of being in the world are dwindling before our eyes.

What characterizes our present predicament is the way in which wilderness (what little remains) comes to represent technology's *other*; when encountered, it reminds us of the pre-technological relationship between the sacred and the profane. The aforementioned shift of balance between nature and culture now takes the form of an inversion: whereas in the mythic experience the erection of a sanctuary established a cosmos and habitat in the chaos of wilderness, the wilderness now appears as a sacred place in the disorientation and distraction of the encompassing technology. As Borgmann urges, in this situation it is a matter of learning again from the ground up what it is to recognize something as other and greater than ourselves; to allow something to be, and remain, itself, sufficient unto itself, rather than laying hands on it and exploiting it for our use and as mediated by technology. In the wilderness we let things be in the fullness of their dimensions, which is only possible when we are open to wilderness' capacity to engage us in the fullness of our capacities, required of us if we in turn are to comprehend and gather wilderness in its extent and depth. By restoring time and space to us, wilderness holds a promise to counter the annihilation of time and space that we saw is essential in technology's disclosure of the world, uprooting everything that comes in its way by robbing it of particularity and rendering it an object of production, exchange and consumption, in short: the logic of anonymity and leveling imposed on commodities everywhere, Nietzsche's eternal recurrence of the same (notwithstanding advertising's tireless insistence on novelty).

Yet the way in which wilderness is experienced by us today cannot help carrying the scars inflicted by centuries of technological exploitation. As Bill McKibben remarks, 'There's nothing there except us'; even the rain has become a subset of human activity (1990: 97). What until recently in human and cultural history was perceived as abundant and all-powerful, now has become so marginalized as to command concern of a wholly novel kind, unknown to our forefathers: the concern that we be contemporaries to, and so witnesses to, their disappearance. Because of the unprecedented nature of this asymmetry of powers between man and nature, inverting the traditional asymmetry, our concern is not over the endangered species, or habitats, or ecosystems as such, as it were; it is rather a concern with them that is inseparable from our concern with the technological powers whose casualties they are. In this sense, our deepest concern is – or should be – not with otherness but with ourselves, or to be precise: with what we, as inhabitants in technological civilization, have become complicit in doing to otherness, threatening that very entity with extinction. 'Simply because it bears our mark doesn't mean we can control it. This new "nature" may not be predictably violent. It won't be predictably anything, and therefore it will take us a very long time to work out our relationship with it, if we ever do' (McKibben 1990: 105). That what since time immemorial used to mark a limit to our powers and aspirations, now exist only at our mercy, is a dramatic turn of events whose full implications we collectively and individually are at a loss to comprehend. So where should we turn for advice and inspiration?

III

Surely cultures we habitually and disdainfully classify as 'primitive' may have something to teach us that we would ignore only at our peril. There is a rich variety of cultures of a low-tech kind whose practices have much to recommend for themselves. For one thing, they tend to leave the stocks intact in the sense of extracting and using up no more than would allow for reproduction. In a word, practices are what we would term 'sustainable', though the word and its present-day connotations would have been unknown to the people in question. That this is so is instructive. Why?

The claim I want to put forward is that there is a limit to the extent to which so-called pre-technological cultures can teach us what to do in our technology-trapped predicament. What limits the example of those technologically 'primitive' or simple cultures from guiding us today, is that the sheer technical inefficiency (comparatively speaking) of their capacities for hunting and harvesting imposes a limit to their capacity to overuse whatever natural resources they pursued in order to sustain themselves as societies, and be it as islands in an ocean of nature's abundance. That said, the role played by technological capacity in leaving large amounts, not to speak of entire regions, out of reach for purposes of human usage, will always interact with input from what we broadly call the ideology of the culture in question: its conception of the world and as part of that of the relationship between man and nature, helping determine the attitudes and values towards non-human creatures such as the animals with which they share habitat.

The point I am driving at can be put like this. It is only when a given culture decides to exploit less of nature than its methods for hunting and for harvesting permit it to do, that we may speak of an autonomous *ethical* point of view in this regard. Limits posited by sheer capacity do not count; they are of an altogether different order. To hunt, fish, extract less than one, by sheer virtue of capacity, *could* do, presupposes an ability and a resolve to limit the degree of exploitation for reasons other than technical and pragmatic ones. 'We could have caught them all, but we decided we didn't want to.'

Such a way of thinking, feeling and acting seems to prevail as long as human needs are not considered to be infinitely elastic, so that getting more entails craving more (consumerism). The latter is a logic where more is regarded as both (technologically) possible and (ethically) desirable. What I term ethics proper – ethical considerations enjoying autonomous authority over what to do and not to do – comes to the fore in this domain only when a limit is decided upon in the following manner: although able to exploit more still, we decline from doing so. Why? Because something is more important than doing (extracting, exploiting) whatever we have a capacity to do. Not everything that we are in fact capable of doing, ought to be done. We deliberately impose a restraint on the deployment of the capacities we command.

That technology cannot be restrained as long as we leave the issue of where to put a limit and declare 'no further' to technology itself is a lesson we are only now beginning to learn, reluctantly and belatedly at that. Present-day technology is capable of intervening into both first and second nature so as to alter – not only exploit – it, as exemplified by biotechnologies, eugenics, and GM. As for the more old-fashioned capacity to hunt and to fish, it is that of overkill, with excess spelling lethal capacity. For example, overfishing is not only technologically possible, it is a reality in many regions. Huge, historic fish stocks have been destroyed within a few years by modern technology, breeding stocks of a basic species like cod wiped out by the huge trawlers on the Canadian Grand Banks, to pick just one instance (Kurlansky 2008: 146).

Regulation, then, is not taken care of by the limits against which our technologically facilitated methods are up. Nature's limits – or academically put, ecological limits – do not speak for themselves. To make a difference to our ways, we must pay attention to them, allow them a voice; to heed them is up to us entirely. Much recent history is about exceeding hitherto technology-immanent limits, having largely to do with efficiency. It is only when humans *can* exploit nature in ways that are damaging nature's capacity for reproducing itself that our relationship with nature is put to the test in an ethical sense. Thus, it would probably be misplaced to admire the American Indians for not wiping out the bison on the prairie: their weapons were too inefficient to allow them to do so. Only when the technology used in hunting is capable of killing more animals than the regeneration of the stock allows for, enters the issue of imposing a brake on one's practices. Here, what I call an ethical and values-based point of view is brought to bear on the question: how much do we take out, and how much do we thereby voluntarily *abstain* from extracting? This reasoning is genuinely ethical in that it addresses the issue of how and where to draw a distinction – to impose a limit – between reasonable and unreasonable exploitation of nature. Again, only in a situation where the technologically *possible* (doable, makeable) exceeds the limits on exploitation postulated from an ethical point of view, does it become evident which of the two conduct-guiding authorities is trump.

Studies of indigenous cosmologies demonstrate how belief in the spirits of game animals restricted overhunting, and how shamanism functioned in the management of natural resources. Everything in the environment is considered to have life and spirit. A traditional conservation ethic can thus be defined as 'the awareness of one's ability to damage natural resources, coupled with a commitment to reduce or eliminate the problem'. For instance, the Cree, like many other groups of North American aboriginal people, consider that it is the animals, not people, who control the success of the hunt. Hunters have certain obligations to fulfil toward the animals, maintaining a respectful relationship that the hunters would only violate to their peril. Following Fikret Berkes (2008: 103), among the Cree respect for the animal is shown in several ways:

- the hunter maintains an attitude of humility when going hunting;
- the animal is approached and killed with respect;
- the animal is carried respectfully to camp;
- offerings are made to the animal;
- the meat is butchered and the meat is consumed according to rules signifying respect;
- the remains of the animal is disposed of properly.

As indicated, the living environment is seen as a community of beings that are supernatural as well as natural. Whereas in Western science it is assumed that humans can control animal populations, in Cree worldview, 'human management' of animals and environment is not possible. Rather, it is animals who control the success of the hunt. The hunter has to show respect to the animals because the hunter is dependent on game. This instils an attitude of humility. The game is not there for the taking. There is no guarantee of a kill; the game must be pursued. Increase in the hunter's success, as he reaches his prime, goes hand-in-hand with the increase of his respect for the animals. The major principle is that everything caught is consumed and there is no waste. It is important that everything is eaten. Killing for fun or recreation or 'sport' without eating is a transgression. What one kills, one keeps for eating (Berkes 2008: 107).

Note that there is no distinction or separation between worldview and ethics. What we as Westerners would identify as the principles and norms contained in the latter, is what is practised, lived, *experienced* by the members of the community: the reciprocal, interdependent relationship between man and nature is a lived relationship, not a theoretical and as such merely postulated one. By acting upon the relationship qua reciprocal and interdependent, the relationship is effectively sustained for the future in these very qualities.

J. Baird Callicott (1989: 189f.) summarizes the wordview in question:

> The implicit overall metaphysic of American Indian culture locates human beings in a larger social, as well as physical, environment. Existence in this larger society places people in an environment in which reciprocal responsibilities and mutual obligations are taken for granted without question or reflection. All creatures, be they elemental, green, finned, winged, or legged, are children of one father and one mother. One blood flows through all; one spirit has divided itself and enlivened all things with a consciousness that is essentially the same. The world around, though immense and overwhelmingly diversified and complex, is bound together through bounds of kinship, mutuality, and reciprocity.

IV

It is often complained that ours is a through and through materialist society; that, as far as ethics and values are concerned, a materialistic mindset in a world of limits is our main problem.

Though widespread, I have always found this notion misleading. It would be more illuminating to say that we are not materialistic enough, meaning by this that we care too little about the world of objects, many of the human-made artifacts, surrounding us. As Jean Baudrillard (1998) observes, whereas a piece of furniture would outlive its users a generation or two ago, techniques such as planned obsolescence has seen to it that the furniture we use is replaced a great number of times within our lifetime. Turnover is increasing everywhere; we throw away things long before they become worn out; we acquire the newest model in full awareness that we will use and keep it for ever shorter periods of time. Among other things, this makes the objects comprising our common, man-made world (Hannah Arendt) less apt than they used to be in their function as carriers of memory, as markers of a common past that we go on actively relating to. So it is not only that present-day society is hell-bent on overkilling, wiping out and destroying an ever-growing number of species in our natural surroundings. It is also that the technology that helps frame our ongoing relationship with the artificial (man-made) world of things plays a huge part in rendering that relationship more transitory and so less conducive to commitment and care than what was the case in the world, say, our grandparents lived in. The demand for replacement of the old with the new in this domain is not at all driven by objective necessity, let alone by what I have identified as ethical valuation and reasoning. This demand for change, for altering whatever was just made and taken into use, is as it were wholly artificial or, if you like, a feature of culture, *our* culture; it wasn't always like this, and it needn't be like this in the future. In light of the above analysis of technology, it is notable that the craving for, nay addiction to, novelty into which young generations are now thoroughly socialized by advertising from early on, has been effectively divorced from the goal of improving quality as measured by longevity and durability, that is, by the product's immanent standards of excellence. That a new model lasts longer than its predecessor, is not the point (it doesn't). Rather, the point is to put something on the market that is meant to last, to be in use, for only a short time; to afford products not meant to last is what is being communicated by buying (wearing, showing off) them, and in this sense it is for the privileged only, at least in the beginning. Customer and so status exclusivity is secured by buying what is designed to be quickly dumped.

Seldom has the perversity of the culture of capitalism been more unashamedly displayed. At precisely the point in history when an unprecedented number of species as well as natural resources like metals and minerals are brought to the brink of extinction as a direct and as such uncontroversial consequence of our technology-driven overexploitation and overkill, this same culture is more intent than ever before on introducing the logic of higher turnover (for example, by way of planned obsolescence) into ever more domains of production and consumption. Present culture's favoured response to the culture-created and now urgent need to save, to extract and use with the utmost care for durability and sustainability, consists in encouraging a consumerist attitude of use and throw away, of wastefulness as the essence of an entire lifestyle.

V

Alluding to Heidegger, Borgmann (1994: 198) asks: 'How are we to recover orientation in the oblivious and distracted era of technology when the great embodiments of meaning, such as the works of art, have lost their focusing power?' Following Heidegger, his answer is that we must uncover the simplicity of things, of the *Dinge* that help make up our world and give meaning to it, especially the way old things maintain for us a physically present connection with the past, even with what went before we were born. This captures the kind of 'materialism' I meant to invoke above, a positive not destructive one. To paraphrase Heidegger's own example: a jug, an earthen vessel from which we pour wine, is such a thing. It teaches us what it is to hold, to offer, to pour, and to give. In its clay, it gathers for us the earth as it does in containing the wine that has grown from the soil. It gathers the sky whose rain and sun are present in the wine. It refreshes and animates us in our mortality. And in the libation it acknowledges and calls on the divinities. In these ways, explains Borgmann, the thing gathers and discloses what Heidegger calls the fourfold, the interplay of the crucial dimensions of earth and sky, mortals and divinities. A thing, in the sense intended, is a focus; to speak of focal things is to emphasize the central point twice.

We appreciate how this Heideggerian depiction of the thing fits with our earlier observations concerning the characteristics of deictic discourse. The vessel serves to focus our attention; it forms part of a practice for which we are grateful; it highlights our embeddedness within an intergenerational, common world; it has already survived our parents' and grandparents' use of it, and we feel responsible for treating it with enough care that it survives our present use and will benefit that of our own offspring. The vessel is a communal, world-disclosing thing. In the many senses and dimensions mentioned here, it forms a counterpart to the use-cum-throw away-cum-replace practices urged by contemporary consumerism, whereby the objects consumed as well as the accelerated pace with which they are so spread to and proliferate within ever-more domains of existence, invading culture (man-made objects) itself in addition to the natural world.

The relevance of Borgmann's Heidegger-inspired perspective at this hour owes much to the contrast it helps highlight for us. It teaches us to do what we are otherwise inclined to abandon, to stop practising altogether, and so collectively forgetting. The orienting force of technology at issue here will come to the fore only as the rule of technology is raised from its anonymity, is disclosed as the orthodoxy that heretofore has been taken for granted and allowed to remain invisible. Borgmann (1994: 199) puts it like this:

> As long as we overlook the tightly patterned character of technology and believe that we live in a world of endlessly open and rich opportunities, as long as we ignore the definite ways in which we, acting technologically, have worked out the promise of technology and remain vaguely enthralled by that promise, so long simple things and practices will seem burdensome, confining, and drab. But if we recognize the central vacuity of advanced technology, that emptiness can become the opening for focal things. It works both ways. When we see a focal concern of ours threatened by technology, our sight for the liabilities of mature technology is sharpened.

However, we should beware of adopting Heidegger's analysis of technology and of the focal thing uncritically. Borgmann is right to remark that Heidegger's suggestion that we have to seek out pre-technological enclaves to encounter focal things is misleading and dispiriting. Technology is here to stay. The task at hand is to exploit the ability of focal things such as the vessel or the shoe, or the centuries-old tree, or what is left of pristine boreal forests, to put the selectivity of the technological way of disclosing – inhabiting – the world into sharp relief so as to make us aware of it: aware that despite its pretension to all-out dominance, that is, to both signifying and exhausting reality as such, technology is nothing more and nothing less than *one* possible, hence limited and limiting, take on the world and our possibilities in it, for it.

That this is so typically emerges out of opposition, of resistance to attempts to reinvigorate focal practices centred around the attention to and respect for focal things. Perhaps the point is best expressed the other way round. In order to feel how focal practices and the things they centre on represent an opposition to, a putting into question of, our otherwise taken for granted outlook on and place within the world, we must have experienced the subtly debilitating character of technology. We must, that is, understand that the peril of technology lies not in this or that of its manifestations, or projects or products, but in *the pervasiveness and consistency of its pattern*. So ubiquitous and taken for granted is this pattern that only by virtue of witnessing its encounter – contrast – with its other will we be able to see it for what it is: not omnipresent and inevitable and coextensive with the world, but instead one among many (still) possible ways of being-in-the-world. As Borgmann succinctly observes, a case-by-case appraisal of technology is inconclusive at best and grossly misleading at worst. There will always be cases where a new device or method is unobjectionable and truly helpful to address genuine human needs; life-saving advances in medicine immediately spring to mind as a case in point. It is when we attempt to take the measure of technological life in its normal totality that we are distressed by its shallowness, not to mention frightened by its powers of destruction. Practices in the sense intended here protect focal things not only from technologicasl subversion but also from human frailty. As we saw above, the ultimately significant things to which we respond and for which we are inspired to care and feel awe in deictic discourse cannot be possessed or controlled. They are in our midst yet reach us from outside, as it were, in that they have the character of a gift, something we are not entitled to dispose of as dictated by the desire of the moment, something for which we carry responsibility, making sure that it survives our contact with it here and now, respecting it as *meant for others*, not only for us. To quote Borgmann (1994: 211): 'Only things that we experience as greater and other than ourselves can move us to judge and change technology in the first place.'

Against this background, the belief shared by the political and economic elite that technology is what will save us appears shockingly naïve. Technology is as much and as deeply part of our current problems as part of their solution. Advances in technology, Jared Diamond (2005: 505) reminds us, 'just increase our ability to do things, which may be either for the better or for the worse. All of our current problems are unintended negative consequences of our existing technology'. To speak of 'all' in this way may be putting it to strongly; it risks simplifying the causes of our predicament, and it seemingly subscribes to technological determinism. Diamond's main point, however, is well taken: there is no reason to think that, at a certain point in our development (say, right now), new technology will cease to cause new (unintended, unforeseeable) problems, that its only impact on problems will consist in making them disappear.

For good or worse, the post-Second World War era was one of trying to convert social into technological problems; its paradigm was that of social engineering. Such paradigms only go, only succeed, so far. Presently, or at least since the publication and worldwide debate over the Brundtland report (1987), the favoured tack has been to convert environmental into technological problems. Politicians urge the scientists to provide a technological fix in the face of the man-made destruction of species and habitats, especially in the form of anthropogenic climate change. But since these problems spring largely from frivolous consumption, would it not be more reasonable to prevent them from arising in the first place than to 'fix' them technologically? In fact, we have seen that it is precisely a combination of the ways in which present society is both technology-driven and consumption-driven that is the motor of the trends tantamount to impending disaster, the combined instance dictating the direction and constraining the safety measures taken and the options considered. What this strategy boils down to is a recipe for disaster, for getting us there even faster, not for averting it.

How we perceive something determines how we are disposed to treat it in our actions. This is the heart of the matter. We have become a lethal force to virtually all forms of life on earth save the human one: to all species save *homo sapiens*. Proceeding along this trajectory is bound to backfire; it is only a question of time before we are the next, and be it among the last, casualties. Whereas other species do not, for the sake of their very existence, depend on us and our flourishing, we depend on theirs. If there is asymmetry here, it is of a kind diametrically opposed to what our arrogance and hubris has led us to believe.

I started out depicting the dangers involved when advanced technology such as the modern computer-based logging vehicle comes to stand between us and the (piece of) nature that we relate to for purposes of exploitation. The manner in which this relationship works out transforms it into a non-relationship in that it is stripped of the various dimensions of sensuous-bodily commerce with the outer reality at hand (i.e. the tree). What drives this is abstraction, both as cause and as effect. We get better at making, say, cellulose, but at the cost of losing sight and touch and smell – losing reality-contact – with the particular materials (forms of life) out of which it is made. For want of direct contact with the living parts of nature whose death and extinction we help cause, we are unable to identify with it, to empathize with it, to care for it. The addressee (man's other) having become abstract and exchangeable, having become sheer quantitative input by way of abstraction, means for it to have become ethically nonexistent, not to be reckoned with as a separate existence with a life and an entitlement to its thriving. This is why revitalizing focal practices, centering on focal things, is an urgent task. Doing so will clearly be to our gain, however much we have come, perversely, to fear it as a loss (of comfort or affluence). As Aristotle argued, the more complex the faculties to whose cultivation we are devoted, the more excellent our life. To engage fully instead of abstractly with non-human life would mean to relate to it in a way conducive to that complexity. And by starting to see, hear, touch, smell and – not least – feel what we do because attentive and receptive to that (those) to whom we do it, we would stand a chance of halting senseless destruction of non-human life before it proves too late.

Bibliography

Anders, Günther, *Die Antiquiertheit des Menschen*. München: C.H. Beck, 1956.

Barnosky, Anthony, *Heatstroke. Nature in an Age of Global Warming*. Washington: Island Press, 2009.

Baudrillard, Jean, *The Consumer Society*. London: Sage, 1998.

Berkes, Fikret, *Sacred Ecology*. New York: Routledge, 2008.

Borgmann, Albert, *Technology and the Character of Contemporary Life*. Chicago: University of Chicago Press, 1994.

Brundtland, Gro Harlem *et al.*, *Our Common Future*. Oxford: Oxford University Press, 1987.

Callicott, J. Baird, *In Defense of the Land Ethic*. New York: SUNY, 1989.

Diamond, Jared, *Collapse*. New York: Penguin, 2005.

Harrison, Robert Pogue, *Forests. The Shadow of Civilization*. Chicago: University of Chicago Press, 1992.

Heidegger, Martin, *The Question Concerning Technology and Other Essays*. New York: Shocken, 1977.

Kurlansky, Mark, *The Last Fish Tale*. London: Jonathan Cape, 2008.

McKibben, Bill, *The End of Nature*. New York: Bloomsbury, 1990.

Meyrowitz, Joshua, *No Sense of Place*. Oxford: Oxford University Press, 1985.

Næss, Arne, *Ecology, Community and Lifestyle*. Cambridge: Cambridge University Press, 1989.

Rosa, Hartmut, *Beschleunigung*. Frankfurt/M.: Suhrkamp, 2005.

4

ECOPHILOSOPHY AND THE CONTEMPORARY ENVIRONMENTAL DEBATE

Karl Georg Høyer

(THIS TEXT WAS ORIGINALLY PUBLISHED IN 2007 IN *SOSIOLOGISK ÅRBOK* (ANNALS OF SOCIOLOGY, NORWAY), NO. 3–4: 2007, PP. 1–35)

Summary

The article is based on a keynote lecture at the Annual Conference of the International Association for Critical Realism, University of Tromsø in 2006. It is about some of the most prominent ecophilosophers of the Nordic countries, and a discussion of the contemporary relevance of their philosophies. Included are the works of three Norwegians, Peter Wessel Zapffe, Arne Næss and Sigmund Kvaløy. The Finnish philosopher covered – Georg Henrik von Wright – was a favoured student of Wittgenstein. In 1948 he took over the Wittgenstein professorship in philosophy at Cambridge. Von Wright was born in 1916 and died just a few years ago at the age of eighty-seven. In his basic philosophical works he is considered to have a central role in twentieth-century philosophy of science. But in his writings in the last twenty years of his life he definitely moved out of the mainstream. He probably would not have agreed to my term *ecophilosopher*. However, these works largely focus on ecological issues. The Swede included is Torsten Hägerstrand. He was a professor of geography at the University of Lund, born in 1916 and died just three years ago, at the age of eighty-eight and active to the very last. He fathered the theory of *time geography*, which has inspired geographers all over the world. Although not a philosopher by profession, much of his writings are philosophical. He considers his time geography to be more of a worldview than a theory of science. Most extensively he has written about ecological problems and their relation to scientific views and developments.

The fundamentals of interdisciplinarity are emphasized in all Nordic ecophilosophy. Both bio- and human ecology focus on wholeness, on complexities, and on the complex inter-relations between the diversity of units, that makes the whole both something more and something else than the individual parts. It is a view that not only gives an epistemological reason, but also an *ontological reason* for interdisciplinarity. There is another relation of importance in this context. The compartmentalized disciplined sciences are in themselves considered to be part of the problem causing ecological crisis. This is very much emphasized by all the Nordic contributors. Arne Næss puts particular focus on the science of economics. He considers this discipline to be an important part of the problem, not a part of the solution. What Arne Næss describes as *shallow ecology* has today gained a dominant position. In environmental politics, and in environmental sciences. However, also in the environmental movement. There were close relations between the ecophilosophers, their ecophilosophy and the environmental organizations until some time in the early 1980s. Today, there are many new organizations. They compete on a market; an environmental market of media attention and technical solutions. They are fully equipped with super-specialists and they have become parts in an environmental career system. Their employees move easily around as if travelling between various stops along an underground subway track of environmental careers, just as in a pipe system, to use the Kvaløy terminology. There is hardly any critique of growth heard anymore. The environmental organizations are mostly participants in the discourses on mainstream economical and technical 'solutions'. As emphasized by von Wright this belongs to a form of *fundamentalism* – or scientism – that entails an uncritical and inflated belief that our techno-scientific rationality can solve all problems.

The American sociologist George Ritzer's theory of grobalization in late modernity is quite in line with the analyses and descriptions carried out by the Nordic ecophilosophers some thirty years ago. And of course grobalization forces expanding and imposing on the local, generate severe ecological imbalances and environmental problems, locally as well as globally. This is not included in the Ritzer theory. Ritzer also launches the thesis of globalization as production of *nothing* and *nothingness. Grobalization* leads to an increasing dominance of nothing in the form of non-places, non-things, non-people and non-service, all at the expense of something on a nothing-something continuum.

In their ecophilosophy both Zapffe and Kvaløy put emphasis on the production of nothing, non-places and non-people. But in late modern sociology, terms like spaces of flows, nothingness and non-places often convey the post-modernistic illusion that this is totally *foot-loose* in relation to nature; that these are processes that also represent nothing in relation to nature. Quite contrary to that, there is nothing that can imply ecological neutrality. Production of nothing definitely leads to something in nature, more severe forms than before. The global flows are fortifying processes of ecological crisis. Non-places are integrated in particularly environmentally harmful societal processes, as *hypertourism, hyperconsumption* and *hypermobility*.

Introduction

This is a presentation of some of the most prominent ecophilosophers of the Nordic countries, and a discussion of the contemporary relevance of their philosophies. We are in Tromsø. At an international conference on critical realism. This is an opportunity to convey some introductory words about the Norwegian philosopher Peter Wessel Zapffe. He was born in Tromsø in 1899. He died in 1990 at the age of ninety-one. He moved away at age thirty, never to resettle in the city again.

But throughout his life he continuously returned to the Troms mountains, in particular to his cabin at Jægervann and the Lyngen Alps. He was a pioneer not only in nature conservation but also in ecophilosophy. Well, actually he did not like the term nature conservation: 'Conservation is definitely an evil in relation to unspoilt nature, just as vaccines are in relation to natural health. But this choice is no longer there. Now, only conservation can save us from Die Norwegishe Apparatlandschaft. I say it in German' (Zapffe, 1969). Today we would perhaps rather say it in Anglo-American.

Shut up

For Zapffe, nature is best off without people; yes, in the long term without the human race at all. Humans can only do harm. To nature, and to themselves. I myself have a lifelong membership in an association he founded in the early 1970s. Actually it is rather a non-association. Only virtual memberships. But with obligations to act according to certain norms. It was named the *Shut Up* association. When you stumble into a particularly splendid piece of nature, or climb to a magnificent mountain top. Shut up. Don't tell anybody. If you do, more people will be attracted and the original qualities will be destroyed. Like the other Norwegian ecophilosophers Arne Næss and Sigmund Kvaløy, Zapffe was a keen mountaineer. He enjoyed the richness and diversity of nature but to him mountain climbing as a human activity was just as meaningless as human life itself. *Once, while climbing the spire of Tromsø cathedral he proclaimed that he could not ascend further by means of the church.* He lived as he taught in all relations. And although married for nearly fifty years to the same woman he chose to remain childless. His name was only to be inherited by a mountain peak he once climbed.

The tragedy of man

His doctoral thesis – *On the Tragic* (1941)[1] – was published during the Second World War. Arne Næss – then a young philosopher – took leave from his activities in the resistance movement to serve as an opponent when Zapffe defended his thesis. According to this philosophy, man is endowed with overdeveloped skills in understanding and self-consciousness. We have become an overdeveloped race in relation to the nature mankind developed from. As long as humans seek meaning and justification in matters like life and death we will continuously meet disappointments. The stronger our craving for answers, the more disappointments encountered. Nature is simply not structured to satisfy such demands.

Nevertheless, it's these demands that define what it means to be human. If people refrained from making them they would also be undermining their humanity. Zapffe was the first Norwegian with a clearly developed ecophilosophy. In a new introduction to his thesis, first published in 1983, he related his works to an understanding of the ecological crisis (Zapffe, 1996). However, the basics were expressed already in his writings from the 1920s, for instance in a poem published in the local paper *Tromsø Stiftstidende* in 1929 (see Zapffe, 1975). His own term is *biosophy*, as he described it in a work on introduction to dramatic literature in 1961 (see also Zapffe, 1975). In this text he also employs the concept of ecology. The biosophy of Zapffe is, by the way, a far cry from the current international biosophy movement and programme.[2]

The entropic process in nature

His biosophy was based on an understanding of the fundamental entropic principles in nature. There are two 'laws' within thermodynamics that are relevant in this context (Høyer, 1979).

The first one is the law of constant quantities of energy and material resources (matter or mass). It tells us that energy, or matter, can neither be created nor destroyed; they can only be transformed from one form to another. If energy in any form disappears in one place, then the same quantity must be found in another form or in another place. The fundamental implications are of course that something can never become nothing, and vice versa, that nothing can never become something. In the language of football: *we can never win, the best we can hope for is a draw*. However, in principle the first law accepts that any transformation process of energy or matter can go both forwards and backwards, even forever. If Newtonian physics doesn't tell us, most of us – including Zapffe – perceive readily that this cannot be how real nature works. This leads us to the second law, which dispels any question of the feasibility of perpetual-motion machines, *perpetuum mobile*. This can most easily be explained in relation to the transformation of energy. When energy is used, that is, transformed from one form to another, it will incessantly move towards an increased degree of disorder, or chaos. Another word for disorder is entropy, thus the name: *the entropy law*. When energy is used it irrevocably loses some, or all, of its ability to be useful, to carry out work.

When used, high-quality energy, with a large working capacity, proceeds continuously towards more inferior energy that has less capacity. Eventually there is complete disorder, or maximum entropy, in the form of finely dispersed heat energy. With football terminology again: *we can not even get a draw, we are doomed to lose*. A real tragedy within the Zapffe philosophy also, because with our human consciousness we are well aware of it.

As emphasized, the same total system disordering process is also applicable to all transformation of material resources. Utilization of material resources is a process both in time and space.

Time cannot be reversed, and the resources cannot be returned to their original state without further generation of entropy. This explains why the renowned Rumanian-American theoretical economist Nicholas Georgescu-Roegen (1971; 1972; 1975a; 1975b) has termed modern economy an entropic process. Economic growth is fuelled by high quality, low entropy sources, but will incessantly produce high entropy disorder. Larger growth, larger production of disorder. The more we invest, the more we are doomed to lose.

Nature itself is on the other hand seemingly a counter-entropic process. High quality, low entropy is continuously built up. However, only through incessant high-quality energy fluxes from the sun. After growth there is always death, and the new life forms can only grow out of the dead ones. In modern society, human encroachments on nature can only take place through destruction of other life forms, in addition to the fortification of the entropic process. And as Zapffe (1975; 1989) has been interpreted by Sigmund Kvaløy: human constructions must also fall, and the grander the constructions built, the more extensive and deafening are the falls.

Industrial Growth Society as a crisis system

The Sigmund Kvaløy term for this human society is the *Industrial Growth Society* (IGS). A later term of his is the Advanced Competitive-Industrial Dominion, abbreviated and with connotations to *ACID* (see for instance Kvaløy, 2000). It is *a system of crisis*; both ecological crisis and human and social crisis. He was considerably inspired by the works of Zapffe. Kvaløy has been – and still is – a very influential ecophilosopher and environmental conservationist, now at an age of seventy-two. The fundamentals of his ecophilosophy were initially developed in the early 1970s (Kvaløy, 1971; 1973a; 1973b; 1975; 1979; 1990; 1991a; 1991b; 2000). He probably would not like my only using the surname Kvaløy. In his later writings he also always uses a second surname Setreng, the name of the mountain farm he inherited from his uncle and where he later has lived. But in this context I'm mostly referring to his earlier works.

On the Industrial Growth Society – or later ACID – he pins the following key characteristics (Kvaløy, 1973a, pp. 147–8):

- It is a society that has as its basis a *continuous growth* in the output of industrial products, one where questions about whether these products are life necessities are not treated as important questions.
- The growth in industrial production is propelled through the mechanism of *competition*, whether it is between persons, industrial corporations, or industrial states. Not to compete implies not to survive.
- IGS's most important industrial resource and means of competitive survival is *applied science*. This – coupled with the propellant of competition – leads to a situation in which the number and variation of ways to utilize nature and humans to promote industrial growth interests is increasing exponentially. The 'front of attack' upon the total ecosystem – including the social system – is widening with increasing speed and the sum total and nature of this attack is continuously growing more difficult to grasp.
- To this last point it must also be added that the mounting competition to get on the market first with new products requires such an increasing number of

scientists and engineers that they become *narrow-viewed super specialists.* These are people who are systematically trained to lose the overview that solutions to the systems crisis demand.

- IGS´s most characteristic tool for social coordination and for repairing social imbalance or disruption is *standardization* and *quantification.* Closely knit and complex social units, established through slow adjustment over the ages to specific and varying local requirements, are cut into pieces, and the pieces refitted into a simplified, standardized, and quantifiable pattern – suitable for centralized, big-unit switchboard control.
- This last phenomenon – that quantification has become the general tool of social problem-solving and steering – has paved the way for the *development of control and manipulation techniques* through the use of computers, communication via electronic networks, and electronics in general. Viewed as a whole, this constitutes a gigantic stride towards shaping humanity into an accessory to machinery, cutting us off from self-control and from the mastery of our destiny in the global environment.

We may recognize these characteristics from many current critical analyses of *globalization.* However, not in the post-modern version of *glocalization,* but in the neo-Marxist version of *grobalization* as the driving force (Ritzer, 2003). Global capitalism is a necessary part of the grobalization processes. Kvaløy (1973a) himself uses the term *techno-capitalism,* also understood as a global force.

Kvaløy (1973a, pp. 148–9) emphasizes that IGS is a society that is building imbalances on all levels – ecologically, socially, economically and mentally – and is doing so at an accelerating pace. The various signs of this imbalance – environmental problems, human stress, violence, etc. – are constantly and by necessity misinterpreted by IGS´s control organs; all problems are primarily understood as being of a technical or economical nature, and solutions are devised that fit such a picture. The 'solutions' themselves thus become factors that strengthen the crisis. Also, social bonds are being torn, meaningless ways of life are spreading, and the resulting unrest and dissatisfaction is compensated for by the *Disneyland effect.* I shall return below to this term, launched by Kvaløy.

Complexity versus complication and living in pipes

This pair of concepts is crucial to Kvaløy's ecophilosophy. This is emphasized by Arne Næss (1985), among others, who claimed that few distinctions in ecophilosophy are so illuminating. Complexity is an organic and diverse coherence, just as it is in ecological complexity. As in system ecology it is *diversity in structure, function and communication.* Complication on the other hand is without this organic coherence, even though a form of diversity may be present. A distinction can be made between artificial and organic diversity, a distinction that is not readily grasped in postmodernistic analyses. Complication is thus an expression for the multi-factored set of entities and relationships that can be analysed in quantitative terms.

Complexity involves operating on *many qualitative levels*, besides the level of complication. Complication was ushered into the world by the human mind. It refers to the special cluster of difficulties and – ultimately – impossibilities inherent in the attempt to grasp and control complexity through quantification. It is this one level of theorizing and organizing that has given man his industrial and technological might, and in the Industrial Growth Society it is replacing all the other levels of relating to our total environment. And the 'Apparatlandschaft', where society primarily masters its total environment through technologies, contributes to the dominance of the level of complication (Kvaløy, 1973a, p. 45).[3]

Kvaløy (1973a, pp. 47–57) compares human life in the 'Apparatlandschaft' with life in a *pipe system*, another of his crucial concepts. The most functional citizens of the IGS society are those with no protruding edges – those who have the slippery shape of simple discs – well suited to slide through the centrally designed pipe system of the modern techno-society. However, if this pipe system should break down, for instance through drastic ecological changes, then the standardized pipe structure and the lack of individual and social complexity would be revealed as extreme vulnerability. The majority – trained to function as simple discs – would then reveal their complete dependence of the tubular guidance of the pipe system to which they have been fitted (p. 153). All of the energy of the 'disc-humans' is used up for one purpose: to manage and communicate *complication.* They move easily within the pipe system, but are lost outside or without it. I have used the concept or metaphor of *life in corridors* in my analyses of the lives of late modern, global *conference tourists* in academia. This I shall return to at the very end of the chapter.

The Disneyland effect

According to Kvaløy the unrest and dissatisfaction being created by complication and life in the pipe system is compensated for through the *Disneyland effect.* He compares it with a continuous life in an amusement park; we are amused by an incessant flickering of symbols and signs and through superficial engagement. This compensatory Disneyland effect absorbs the need humans have for development of physical and psychological complexity, needs we are all born with. In this way people are obstructed from developing their powers of life, which is also the case for human society on the whole. This, however, need not immediately imply a similar reduction in pure survival strength, neither on an individual nor a societal level. Here is another pair of concepts he applies: *life strength* versus *survival strength* (Kvaløy, 1973a, pp. 57–60).

Zapffe (1975) conveys a similar understanding. He defines several levels of what he terms *interest fronts*: *biological, social, autotelic* and *metaphysical.* Autotelic refers to whatever is rewarding in itself. Within the autotelic area, offers are especially extensive in the modern development of human societies; non-committal amusement is produced in quantities far surpassing the human abilities to receive them, physically or psychologically. This development unrelentingly creates more and more standardized, un-personal individuals and what he terms *nothing-humans.*

I'll return later to these concepts – Disneyland effect and nothing-humans – in relation to the late modern globalization theories developed by the American sociologist George Ritzer. It refers to his works on *enchanting a disenchanted world* and the *globalization of nothing* (Ritzer, 1998; 1999; 2004a; 2004b).

From changes in space to changes in time

In many of his later works, Sigmund Kvaløy (1985a; 1985b; 2004) has given the concept of time a prime position in his ecophilosophy, while he is quite critical of the centrality of the space concept in ecological sciences. He has been highly inspired by the works of the philosopher Henri Bergson (1859–1941), whom he actually considers to be the first European ecophilosopher. Bergson defines two distinctly different forms of time; *la duree* and *le temps*. In our real lives we all live in *la duree*, a qualitative form of time, where each moment is something concrete and special, different from the moments that follow. This is quite another form of time than *le temps*, which is the quantified clock time belonging to an abstract world of mathematics and physics, and where time is subdivided into hours, minutes, seconds, and now even micro- and nano-seconds. Bergson considers the quantified 'le temps' to be a product of Cartesian philosophy; and claims that it is a mistake if one confuses 'le temps' with the real-world 'la duree'. Such confusion is intimately related to what he terms *l'ivresse mechanistic*, or the mechanistic drunkenness. Today the British *Barbara Adam* (1996) is a social theorist who works along similar lines, and who emphasizes the close relations between the dominance of a mechanistic concept of time and the ecological crisis.

Kvaløy (1985b, pp. 25–33) outlines his two concepts of time according to the difference between *complexity* and *complication*. To him complex time on the one hand, and complicated time on the other, relate to very different forms of human cultures. Complicated, quantified time is at the very basis of his concept of industrial growth society. While nature, and all humans within it, live in complex time. With its Greek heritage, European history has given us two divergent kinds of movement. One is mathematically defined, quantitative and lacks qualitative aspects. But it is a false form of movement, as it is based on repetitiveness – nothing happens, nothing changes.

The other form of movement is complex, qualitatively changing and totally outside the realm of mathematics. Nature is of course complex; it has a qualitative diversity, and is a stream of events that never repeat themselves. Thus, it is principally different from any machine structure. A basic characteristic of machines is that they can be started, stopped, and restarted according to standardized procedures. These are the only things in the universe that we can control. But we cannot control nature. Any efforts to do that lead to catastrophe. We also constantly experience that we cannot have perfect control over machines. Complexity is always involved in human–machine relations through physical wear-and-tear and when machines encounter the forces of nature.

Kvaløy (1985b, p. 25) emphasizes that some human cultures have demonstrated wisdom both in relation to nature and to their machines, or technologies. Our originally European culture is vastly unwise in both these relations, and is the first human culture in history that has based itself solely on the machine-attitude towards nature. As this industrial growth culture in its current Anglo-American phase has become globalized, Gaia, our living global organism, has become so seriously injured that the damage can only be surpassed by cosmic catastrophes, such as collisions with comets.

From Wittgenstein to von Wright

We are all acquainted with the main philosophy of Wittgenstein and his influence on the so-called language turn in late modern social sciences. Language is according to Wittgenstein decisive both for how we think and how the world is expressed. One road towards understanding of the world lies in analyses of language; as the world is mirrored in the language. But he denotes feelings, beliefs, and love as *the mysterious*. Language cannot accurately reflect this.

But, and there is definitely a 'but' here, Wittgenstein – perhaps we should denote him the second – did most of his writings in a small, fairly isolated house outside Skjolden, a small rural community in Sogn in Norway. He wrote mostly in isolation and he hardly spoke to anybody in the local area. He was very far from the total involvement of Zapffe and Kvaløy. But in a letter to his friend Moore in Cambridge he still writes: 'I cannot think that I could have worked like I do here any other place. It comes from the calmness, and perhaps from the marvellous landscape. A landscape filled with quiet seriousness.' He was obviously not using linguistic communication in relation to his surroundings. But language was not the prime inspiration for his philosophy; it was the mysterious world in the form of the materially present landscape.

The Finnish philosopher Georg Henrik von Wright was a favoured student of Wittgenstein. In 1948 he took over the Wittgenstein professorship in philosophy at Cambridge. Von Wright was born in 1916 and died just a few years ago at an age of eighty-seven. He is globally recognized as an eminent philosopher, besides the Dane Søren Kierkegaard the only Nordic philosopher belonging to the world's 'top twenty'. In his basic philosophical works he is considered to have a central role in twentieth-century philosophy of science. But in his writings in the last twenty years of his life he definitely moved out of the mainstream. The Norwegian philosopher Jon Hellesnes (1996) claims that von Wright's British and American admirers would be shocked if they discovered these later works. And von Wright has definitely distanced himself from the Wittgenstein language philosophy. He probably would not agree to my term *ecophilosopher*. However, these works largely focus on ecological issues (von Wright, 1986; 1993; 1994; 1996).

The deep historical roots to the ecological crisis and the impossibility theorem of sustainable growth

Von Wright is gravely pessimistic about the prospects of developing sufficiently deep solutions to the ecological crisis. He designates his position as *provocative pessimism*, a term he borrows from the late Norwegian ecological socialist Hartvig Sætra. He thinks this is indeed a crisis with very deep historical roots, mainly described as *a combination* of two factors. The first is the special form of rationality and reason that is expressed through the exact natural sciences as they have been developed since the early works of Descartes and Galileo in particular. In all its dominance he considers this a form of *fundamentalism* – or *scientism* – that implies an uncritical and inflated belief that our technical-scientific rationality can solve all problems. *Zweckrationalität*, to use the Max Weber term. Under the sway of this fundamentalism, questions that fall outside the defined rationality domain can only be subject to 'feelings' and unfounded claims, rather than to rational deliberations. This is nearly congruent with the Kvaløy analysis. The second key historical root factor is the view that we humans have the right to govern nature and do whatever we want with it. This stems from a cultural heritage which, via Christianity, hails from the Jewish creation myth in the Old Testament. Not all cultures in human history share this perspective on the relation between man and nature (von Wright, 1996).

Von Wright describes (1986) two principally incongruous types of relations between man and nature. The first entails seeing the larger whole in nature, a realm where we belong and have to organize our lives according to nature's principles. The second is to consider nature primarily as something putting up resistance, which mankind needs to fight; either in the form of protection against an intruding nature (wild animals), or by taming it into submission (technologies). The second position is of course totally dominant today but von Wright emphasizes the need to come to terms with nature through the first approach. In this context he outlines four different *ethical issues* that must be resolved in relations between humans and nature (1994):

1. Our responsibilities in relation to our *own bodies* and the bodies of other species close to us. Genetic- and bio-technology are today causing particularly acute problems.
2. Our responsibilities in relation to these *other species*, among others connected to the issue of *animal rights*. The question of transplantation of animal organs to humans is an acute problem area today.
3. Our responsibilities in relation to *future human generations*. An acute problem is our current types and levels of nature utilization. These jeopardize our descendents' opportunities to live their lives in ways we consider acceptable and human.
4. Our responsibilities in relation to *nature as a whole*. Acute problems relate to the *inherent* rights of landscapes and of the systems and diversities of plant and animal species.

These are generally included in the *deep ecology* principles of the Norwegian ecophilosopher Arne Næss, as they will be outlined later on.

Von Wright considers the *sustainable development* concept as it was launched with the UN Brundtland report in 1987 to be highly problematic. He of course agrees with one main component of the concept: that the global ecological condition is so endangered that we must reduce both further utilization of natural resources and environmental damage to a minimum. But according to von Wright, the Brundtland concept has another main component as well; the growth in industrial countries can carry on, the material standards of living in these countries shall not be reduced, and people shall not be forced to refrain from the thousands of new *toy products* that the advanced technological industry spews out in ever increasing quantities to geographically expanding markets. As I mentioned, Georg Ritzer's term is *Grobalization*. Combining the two components into one sustainable development is highly unrealistic – or adventurous – according to von Wright. It's more realistic to assert that the human race, through its expansion as species and utilization of nature, proves its lack of ability as a biological species, and thereby dooms itself to ruin and extinction (von Wright, 1993). In this we can of course recognize the human tragedy perspectives of Zapffe.

Geography in both time and space

I'd like to include a Swede in this text – Torsten Hägerstrand. He was a professor of geography at the University of Lund, born in 1916 and died just three years ago, at an age of eighty-eight and active to the very last. He fathered the theory of *time geography*, which has inspired geographers all over the world. A confirmation of his international position is that he achieved five honorary doctorates at British, American and Norwegian universities, and several international books have been published to honour him (Hägerstrand, 1981; 1987; 1988; 1991). Although not a philosopher by profession, much of his writings are philosophical. He considers his time geography to be more of a worldview than a theory of science (Hägerstrand, 1991). Many of his publications are essays. All are interdisciplinary, utilizing contributions and knowledge from an array of scientific disciplines. He has extensively written about ecological problems and their relation to scientific views and developments; ecological challenges were his main focal point during the last thirty years of his life.

Landscape is his crucial concept. Hägerstrand asserts that all science throughout history has basically had two points of departure; the celestial world and the landscape. With our thoughts and actions we humans are part of the landscape together with other phenomena. In this sense, a person is a 'thing,' that is, corporeally delimited and occupying space as a package of relations. However, humans have now become a geological force in the landscape and, aided by technologies, move vast quantities of materials from place to place. The idea of a recycling society is in this context not more than an idealized figure of thought; it is a notion that represses the fact that physical material is moved between physical places and not between concepts. It is not possible to re-create that which has already planted its footprints in the landscape; time is of course irreversible (Hägerstrand, 1991, p. 12). In this context he is closely in line with the entropic principles in nature I have formerly described, and which so much inspired the ecophilosophy of Zapffe.

A contribution from science should be to analyse and show how events follow events, and how all organisms and 'things' are born and die through some form of life-lines, or so-called *trajectories*. According to Hägerstrand everything that turns up in the landscape has its own biography or history. And there are a lot of things that we cannot see or experience directly with our senses, but which still have a physical reality. He considers it important to make the whole range of seemingly invisible social and economical processes physically perceptible so we can grasp their total impacts in the landscape. This requires that the scientific community takes responsibility for conducting synthetic analyses, where important connections in time and space are to be addressed and the internal and external side of all events should be visualized (Hägerstrand, 1991, pp. 13–15).

He acknowledges the seriousness of global ecological problems, but warns against an understanding that detaches these from their local connections. Global problems can be understood as the sum-effects of 'the tyranny of small decisions', or 'the tyranny of small technologies' (1991, p. 193). Seemingly innocent actions on the micro-level lead to extensive effects in macro as they become multiplied and distributed; our highly developed abilities to multiply are not utilized to analyse and manage the future sum-effects of all individual decisions.

Hägerstrand views motoring as an evident area in this context. Even if new technologies remove the polluting emissions of individual cars, the growth in motoring will accumulatively lead to serious problems. Such levels and forms of mobility cannot be made an option in all parts of the world without severe ecological problems, whatever source of energy they should use.

The *global flows* are important. However, they are just as much connected to place and locality as our historically earlier local flows, both regarding where they come from and where they end up, and whether they are in the form of decisions and actions or in the form of material sources and sinks. In this, social sciences should contest the globalization thesis regarding a development from 'spaces of places' to 'spaces of flows' (Castells, 1996; Urry, 2003). There are no spaces of flows without local preconditions and footprints. Decisions and actions are taking place locally, just as their material mobilization and footprints are local.

Hägerstrand believes a priority task for science – necessarily an *interdisciplinary science* – when confronted with the most serious global ecological problems is to achieve understanding of the local–global interconnections in nature and society and to unveil the many material in-betweens in these interconnections. Certainly, such science should be rooted to the ground and materially concrete (1991, p. 192, p. 196).

Science as a problem

Hägerstrand's views on science as a problem are closely in line with the viewpoints expressed by Kvaløy and von Wright, also supported by Arne Næss (1975b) as we shall see later. Hägerstrand emphasizes that the technologically most successful sciences – physics, chemistry and molecular biology – have been developed according to a Newtonian program, with roots in the rule originally formulated by Descartes, 'subdivide any problem in as many parts that are possible and required to solve it in the best way'. The underlying precondition is that the whole equals the sum of its parts (Hägerstrand, 1991, p. 194).

This Cartesian rule has been followed in the sense that science has identified and studied ever smaller pieces of the material world; right down to the sub-atomic level, where one cannot use terms from the macro world to concretely express what really exists and occurs. This has resulted in an incredible amount of highly specialized disciplines, making it difficult to get an overview. The proponents and participants of these disciplines meet insurmountable problems when trying to combine their results with others, hampering not only the understanding of simple, idealized systems, but also the complex systems in the real world we all live in (1991, p. 194). The micro-orientation in natural sciences is particularly problematic when it is transferred to applied science, leading to new technologies. Such technologies are well furnished regarding their inner structure and their most immediate effects, especially under laboratory conditions. Hägerstrand claims that one could consider factories and other similar closed manufacturing units as large laboratories. On the other hand, the wider-ranging direct effects, and more so the extensive indirect effects, on society and ecology are rarely understood in terms that make relevant impact assessments possible. He refers to the American system ecologist Eugene P. Odum who emphasizes the problems with 'increasing specialization and the preoccupation with laboratory study. The open systems of realworld environments cannot be enclosed in glass tubes or laboratory walls'.

Hägerstrand thinks gene technologies typically exemplify such a fundamental lack of balance between highly efficient micro-knowledge and vague or largely insufficient understanding in macro. The release of genetically manipulated organisms could start a chain reaction that we can neither understand nor control. This chain reaction will not be in the laboratory but in our real world (1991, p. 194).

A basic Hägerstrand concept is *mobility* – an expression of movement. Essentially, there is nothing that can be deemed a neutral movement. There is no social neutrality. And there is no physical or ecological neutrality. Illusions of such movement neutralities have a long history. Almost every traditional culture has its fables and fairy tales about traversing long distances using flying carpets, seven-mile boots, etc. No limits, no changes, and no impacts. Such folklore is also well known from my own country – Norway. As far-fetched as these tales are, they actually provide quite an appropriate description of the current global developments of mobility. Many fail to realize there are clear ecological limits to the extent of movements (Hägerstrand, 1993; Høyer and Aall, 2005).

This separation between movement and change has a place in the very foundation of *modernization*. The scholars of Antiquity did not draw the line in this way. Aristotle's physics and biology make a good example. To him, movement and change were one and the same; all forms of changes were understood as forms of movement. Not only did this apply to movements as such, but also to growth and even to changes in color in nature. Galileo and Newton contradicted this view with their physics. Their programme could not be fulfilled if all changes were seen as based on one common principle. They had to make a distinct category of physical movements between places and points. This came to be an important part of the modernist worldview, and the prevailing process of modernization which we still are part of, even if we now call it late modernity (Hägerstrand, 1993; Høyer and Aall, 2005). According to Hägerstrand, *all social sciences* are largely part of the same broad development patterns as the natural sciences – in particular specialization and the breaking of wholeness into pieces. In this context they long ago reached a limit where they can no longer increase our knowledge. He emphasizes the need to link them to the earth, to ecology and to technology; a form of *ecologization* must take place, both of economy, political science, and sociology (1991, p. 17).

Deep ecology

The Norwegian professor of philosophy Arne Næss is world renowned for his writings on deep ecology, a term first coined in 1974 (Næss, 1974). He's had an extensive impact on international discourse regarding ecophilosophy and environmental ethics. Næss was born in 1912, and at age ninety-five is still active. Like compatriots Zapffe and Kvaløy, he has been a keen mountaineer and active in nature conservation. All three are, by the way, characterized by their total involvement in relation to nature and nature conservation, and their diversities in forms of communication. Næss has also always supported non-violent – albeit sometimes lawbreaking – demonstrations and direct actions against large Norwegian hydropower development schemes, protests in which Kvaløy has been a key player and organizer. And Næss very early introduced *Gandhian* non-violence to the ecological movement. (Some of the more important writings by Arne Næss in ecophilosophy are: Næss, 1972; 1974; 1975a; 1989; 1993a; 1993b; 1997; 1999.) He considers deep ecology to be a movement encouraging people to thoroughly question the fundamental presuppositions underlying the dominant economic approach of Western society in terms of value priorities, philosophy and religion. He presents eight points defining the *deep ecology* perspective (Næss, 1993a, p. 412):

1. The well-being and flourishing of human and non-human Life on Earth have value in themselves. These values are independent of the usefulness of the non-human world for human purposes. This can be termed the *inherent value of nature.*
2. Richness and diversity of life forms contribute to the realization of these values and are themselves intrinsic values. Humans have no right to reduce this richness and diversity except to satisfy *vital needs.*
3. The flourishing of human life and cultures is compatible with a substantially smaller human population. The flourishing of non-human life *requires a smaller human population.*
4. Present human interference with the non-human world is *excessive*, and the situation is *rapidly decaying.*

5. Policies must therefore be changed. These policies affect basic economic, technological and ideological structures. The resulting state of affairs will be *deeply different* from the present.
6. The ideological change will be mainly that of appreciating *life quality* – dwelling in situations of inherent value – rather than adhering to an increasingly higher standard of living.
7. There will be a profound awareness of the difference between bigness and greatness.
8. Those who subscribe to the foregoing points have an *obligation* directly or indirectly to try to implement the necessary changes.

This could most fundamentally be interpreted as a programme for humans to attain *peace with nature*. Næss considers these to be points that many can agree on despite various cultures and religious or philosophical positions, actually comprising his first superior level of fundamental premises. He thinks Buddhist, Christian and Philosophical (Spinozist or Whiteheadian) positions contain ultimate premises that could lead to accepting the second level of the eight common principles of deep ecology. He defines four levels in a derivational system (1993a, p. 413):

Level 1: Ultimate religious or philosophical premises. Many various positions are possible.

Level 2: The common eight-point deep ecology platform. Only these eight points, and common to everybody.

Level 3: General Normative Consequences and 'Factual' Hypotheses. The possibilities widen again.

Level 4: Particular rules on decisions adapted for particular situations. The possible alternatives widen again, dependent on the actual societal conditions.

Two Norwegian ecophilosophies

To describe this derivational system Næss uses the *apron* as a metaphor. It is wide at the top (level 1), narrow at the middle (level 2), then widening again towards the bottom (from level 3 to 4). However, he emphasizes that nature conservation strategies are more eagerly implemented by people who love what they are conserving and who are convinced that what they love is inherently loveable; 'Such lovers will not want to hide their attitudes and values, rather they will increasingly give voice to them in public. They possess a genuine ethics of conservation, not merely a tactically useful instrument for human survival' (1993a, p. 412). Love of nature is thus – if not a prerequisite – considered to be important in attaining peace with nature. *Co-presence with nature* has for all the three Norwegian philosophers – Zapffe, Kvaløy, Næss – been a self-evident part of this.

Kvaløy (1985; 2002; 2004) still strongly emphasizes that there are two highly different ecophilosophies in Norway. One is the Næss *ecosophy T*, which to him is only one among many ecosophies, in accordance with the apron metaphor. Ecophilosophy is to Næss a purely descriptive field of study, while the many ecosophies all express definite positions, specific points of view (Næss, 1989). This is denied by Kvaløy, who claims that ecophilosophy implies total involvement in every way, also in political movements and conflicts. He thereby also completely denies the descriptive-normative dichotomy applied by Næss. Kvaløy criticizes Næss further for his thinking in terms of political harmony and for not clearly expressing the necessity of deep political conflicts, not only deep ecology. And in this context it is possible to see that there have been some changes in the Næss ecosophy from his earlier works (see: Næss, 1972). However, even if his later works do not include a clearly formulated basic value, only manifold value options, they still have deep political implications. When he writes about ecopolitics *within* ecosophy, he emphasizes the need for societal self-determination, self-reliance and the real realization of local communities, hardly harmonic viewpoints in our time. This is also included in the American edition of his main book *Ecology, Community and Lifestyle* (Næss, 1989).

The changes may be more prominent in the Kvaløy ecophilosophy. There is definitely a *Kvaløy I* and a *Kvaløy (Setreng) II.* As already outlined, Kvaløy II puts the concept of time – organic time – at the centre of ecophilosophy. Everything is process in time, and everything is changing with time (2004). Individuals, society and nature itself. He emphasizes the need to do away with all spatially related concepts, even with the concept of space altogether. Earlier ecophilosophical concepts as ecological equilibrium, ecological structures and steady-state society are among those ready for the dustbin. In this context he even considers the Næss term *gestalt* to be too spatial. However, *if everything is time and change, there is the danger that it can become nothing.* It is a philosophy that can end up on the nothingness end of the something-nothing continuum described later. I have earlier talked about the 'Global flows' theories of Castells and Urry; described as an historical development from 'spaces of places' to 'spaces of flows' (Castells, 1996; Urry, 2003). In these theories nature is largely non-existent. The global flows – all processes of change – are seemingly without material anchorage, and without ecological footprints in place and space. This is of course an illusion; there are no global flows without local preconditions and footprints, necessarily in space. In his ecophilosophy – the time-space geography – the Swede Hägerstrand manages to link changes in time and changes in space together. And we have lately based on this sort of deep ecology developed radical analytical tools as *ecological footprint* and *ecological space*, both of course spatial terms.

Shallow ecology

The common emphasis on the need for love of nature among all the Norwegian ecophilosophers leads us to the shallow ecology concept developed by Arne Næss as a dichotomy to the deep ecology. He uses the debates about various environmental problems to clarify the distinctions between deep and shallow positions. Such topics of discourse are: pollution, resources, population, appropriate technology, environmental ethics, and education and science (Næss, 1993a, pp. 414–16).

The shallow approach in relation to *pollution* implies that technology is the solution. Laws limit permissible pollution, so highly polluting industries tend to be exported to developing countries with less rigorous laws. As regards *resources*, the shallow approach has its emphasis only upon resources for humans, especially for the present generation in affluent societies. In this view, the resources of the earth belong to those who have the technology to exploit them. There is confidence that resources will not be depleted because, as they get rarer, a high market price will conserve them, and substitutes will be found through technological progress. In the case of *population* the shallow approach sees the threat of overpopulation mainly as a problem for developing countries. One condones, or even applauds, population increases in one's own country for short-sighted economical reasons. With a shallow approach in *environmental ethics*, landscapes, ecosystems, rivers and other whole entities of nature are conceptually chopped into fragments, thus disregarding larger units and comprehensive gestalts. These fragments are regarded as the properties and resources of individuals, organizations or states.

Regarding *appropriate technology* a shallow approach implies that industrialization of the Western type is held to be the goal of developing countries. The universal adoption of Western technology is deemed compatible with cultural diversity. In *education and science* the shallow approach implies that the degradation of the environment and resource depletion necessitates the training of more and more 'experts' who can provide advice on how to continue combining economic growth with maintenance of a healthy environment. We are likely to need an increasingly dominating and manipulative technology to 'manage the planet' when global economic growth makes further environmental degradation inevitable.

Nordic ecophilosophy and late modernity: some issues of relevance in politics and social science

Ecological enlightenment?

The German Ulrich Beck is a prominent sociologist who applies the *ecology* term very much the same way as the works within Nordic ecophilosophy. After publishing his original book on the risk society in 1986, he has published books such as *Ecological Enlightenment* (1995a) and *Ecological Politics in an Age of Risk* (1995b). The subtitle to his book on the risk society was: *Towards a New Modernity* (Beck, 1986). These works have later inspired several sociologists, to the extent that we can talk of a widely applied new theoretical outlook in international sociology. Anthony Giddens is another contributor to this theoretical field. In relation to the long historical lines in sociology, where nature largely has been non-present, this is of course a most interesting situation. Particularly because of the crucial importance ecological issues are given in the theorizing on dominant aspects and development patterns in the late modernity phase. In this context, Beck is even writing about the successes of the ecological movement.

A major Beck thesis is that a multiple disjunction separates the risks of early modernization from the hazards in the technologically advanced society in late modernity. Hazards and risks take a more dominant position in late modernity than in the earlier phases of modernization. While the production and distribution of goods dominated before, this has now become the case for production and distribution of the bad – ecological risks and hazards. And they cannot be externalized; on the contrary they are continuously internalized in the sense that they are essential parts of dominating development processes.

The hazards of late modernity cannot be delimited spatially, temporally or socially; they encompass nation states, military alliances and all social classes. By their very nature they present wholly new kinds of challenges to the institutions designed for their control. Established rules of attribution and liability – causality and guilt – break down. This means that their careful application to research and jurisdiction has the contrary effect: the hazards increase and their anonymity is legitimized (Beck, 1995b, Introduction). Here we can recognize analyses in the Nordic ecophilosophy.

A claim made by Beck is that *ecological enlightenment* is an integral part of development in late modernity. It is included in his theory on reflexive modernization. He may be right about this. But seen in relation to the early historical processes of realizing the basics of '*Du contrat social*', a component of the modernist enlightenment project, I would claim that we are currently even farther from realizing the fundamentals of '*Du contrat ecology*'[4] than we were in the 1970s.

There is another present trend in this context that troubles me almost as much. I'm referring to some rather dominant leanings within *social constructivism*, in social sciences, and especially in sociology. Yes, even within the discipline denoted *environmental sociology*. For instance; the UN Panel on Climate Change is studied as a social construct. All their ideas are social constructions. Then the climate change and its related ecological crisis problems are also considered to be social constructs. And while these social scientists are putting a lot of effort into their discourse analyses of the climate panel and climate change issues, glaciers are melting, slides and avalanches severely increase, hurricanes occur more often and are stronger than before, large new areas are becoming subject to desertification, and thousands – or soon millions – of poor people are forced into mobility as the ecological refugees of our time.

In the aftermath of the 1986 Chernobyl accident we have seen a similar trend in social sciences. The official limits on radioactivity levels in meats and other food products have been studied as social constructs. And while the social scientists carry out their discourse analyses in order to grasp the social processes that constructed such health limits, radioactivity is still heavily accumulated in the food chain: lichen–reindeer–Sami people, and thousands of children in Ukraine, Belarus and Russia suffer from leukaemia and several other diseases related to radioactivity. Should we really consider these trends in social science as proofs of a broader social process of *ecological enlightenment*, as the claim has been raised by Ulrich Beck?

I would like to emphasize the following; one thing is for certain: the planet is not growing. And undeniably, the global economy is continuously growing, even at an exponential rate. Neither of these two certainties can be social constructs. Thus it should be just as certain that something growing within the constraints and basis of something that doesn't grow is destined to lead to deep conflicts, at one time or another. This is one of the very basics of the ecological crisis.

Interdisciplinarity and total involvement

The fundamentals of interdisciplinarity are emphasized in all Nordic ecophilosophy. Both bio- and human ecology focus on wholeness, on complexities, and on the complex inter-relations between the diversity of units, that makes the whole both something more and something else than the individual parts. Interdisciplinarity is considered a basic condition for the study and understanding of these complexities. It is a view that not only gives an epistemological reason, but also an *ontological reason* for interdisciplinarity.

There is another relation of importance in this context. The compartmentalized disciplinary (disciplined) sciences are in themselves considered to be part of the problem causing ecological crisis. This is very much emphasized by all the Nordic contributors. In his works from the early 1970s Arne Næss (1972; 1974) put particular focus on the science of economics. He considers this discipline to be an important part of the problem, not a part of the solution. And I must emphasize that his analysis was carried out many years before neo-liberal economies took over.

Two long historical trends should be considered important in this context. The first is the type of reductionism, which implies the study of small units in isolation, and where these units through history have been continuously subdivided and become smaller and smaller. The other is this type of reductionist science adapted to applied science as the prime means of development of technologies. These technologies are crucial forces behind the ecological crisis in two ways. First, by directly causing the most serious ecological problems. Our history of studying atoms, molecules, and now the study of genes, have all led to the most serious ecological problems when put into an applied science context. From atomic energy, from the chemical industry and most recently from the genetic engineering industry. This history shows that when we manipulate the smallest units in nature to produce new technologies, the ensuing effects on humanity and nature are vastly extensive. The second way these technologies increase the ecological crisis involves their continuous reinforcement of the 'Apparatlandschaft' between humans and nature, thereby undermining the opportunities for co-presence and love of nature.

The principle of *repetitiveness* is crucial in these types of sciences, and in their related technologies. But such repetitiveness does only take place in *closed systems* manipulated by humans, as in laboratories. We will never find it in nature, as strongly emphasized by both Kvaløy and Hägerstrand. In nature there are always *open, complex systems*, continuously changing with time. Many of our most serious ecological problems can be explained this way; technologies, their products and substances, developed and tested in closed systems under preconditions and *illusions of repetitiveness*, are released into the real nature of open systems lacking repetitiveness. We are always taken by surprise when we experience new, unexpected ecological effects. But it is not surprising at all; under these conditions such effects will necessarily turn up all the time.

Interdisciplinarity is important in two contexts here. Firstly, in order to analyse and comprehend the complex, open systems in nature. And secondly, and in the long term much more important, in order to develop alternative technologies that are adaptive and can function ecologically and benignly in interfaces with real nature and its open systems. In both contexts there is the need to understand humans as parts of nature and with complexities in their interrelations with the technologies they form.

As formerly emphasized, Nordic ecophilosophy is also a philosophy of *total involvement*. And there is a relation between interdisciplinarity and total involvement. Kvaløy stresses the importance of educating what he terms *super-amateursand generalists – trained to combine very many different forms of knowledge*, in contrast to the super-specialists moving and living in his pipe system. And the ecophilosophers all lived as they taught. They involved and communicated on very many levels. Mountaineering, non-violent direct actions to preserve nature, communication through drawings, poetry, essays and humour. The Norwegian Zapffe is, besides his philosophy of the tragic, most known for his written collections of popular comical stories and essays, especially conveyed in local dialects from his part of the country – Troms. Zapffe was a lawyer before becoming a philosopher; and he presented his final thesis for the law exam in poetry!

Shallow ecology and the contemporary ecological movement

The Næss *shallow ecology* has gained a dominant position in environmental politics, and in environmental sciences. However, also in the environmental movement. This reflects on the Norwegian situation.[5] There were close relations between the ecophilosophers, their ecophilosophy and the environmental organizations until some time in the early 1980s. There was even a contemporary organization founded on Norwegian ecophilosophy, with a specific working group in ecophilosophy, and where Kvaløy in particular had a leading role. In short the organization was called (snm). Small letters in brackets. To remind the members – and everybody else – that an organization is never the goal, it is the means.

Today all environmental organizations use large letters, and no brackets. The organizations are up front; they are the goal. And in some cases it also looks as if their leaders have become the goal. There are many new organizations. They compete in a market; an environmental market of media attention and technical solutions. They are fully equipped with super-specialists and they have become parts in an environmental career system. One year these environmental super-specialists are employed in the environmental organization, the next in the Ministry of Environment, in environmental research, or in a private consultancy firm. Some of these organizations are set up and function as private consultancies. Their employees move easily around as if travelling between various stops along an underground subway track of environmental careers, just as in a pipe system, to use the Kvaløy terminology.

There is hardly any critique of growth heard anymore. The environmental organizations are mostly participants in the discourses on mainstream economical and technical 'solutions'. There are hardly any differences between the 'solutions' proposed by the environmental groups and those that are part of the governmental policies. It is mostly a matter of patience in relation to impatience. But they are all actors on the same two fields: the field of *economism* and the field of *technologism*. Both imply that while solutions in these fields might function in certain minor cases, this is grossly generalized to imply that they have solutions to all ecological problems. As emphasized by von Wright this belongs to a form of *fundamentalism* – or scientism – that entails an uncritical and inflated belief that our techno-scientific rationality can solve all problems. We have a thirty-year history of ecological solutions developed under such an umbrella of understanding. It is a history where the solutions themselves have become the problems, often more serious than before, or implying that the problems only have been moved, from one place to another, from one sector to another, or from one topic to another. This history, the sad history of ecological solutions, is swept well under the carpets by all the contemporary players on the market of environmental economism and technologism.

Is this really an expression of ecological enlightenment? And it may even be the time to raise a most critical question: to what extent can we call these organizations real NGOs (= ngos!); are they not really PGOs, Pro-Grobalization/- Governmental Organizations, rather than representatives of the weakest parts of all – nature and its fragile ecology?

Grobalization

The aforementioned American sociologist George Ritzer (2004a; 2004b) has launched the concept *grobalization* in contrast to the *glocalization* theses within globalization theory. Grobalization is the process in which growth imperatives – e.g. the need to increase sales and profits from one year to the next in order to keep stock prices high and growing – push organizations and nations to expand globally and to impose themselves on the local. He says grobalization consists of three major subprocesses: capitalism, McDonaldization and Americanization. His is a concept greatly at odds with the dominant post-modernistic theses of glocalization. Ritzer thus underlines that there is a gulf between those who emphasize the increasing grobal influence of capitalistic, Americanized and McDonaldized interests, and those who see the world growing increasingly pluralistic and indeterminate.

Largely as antitheses to the glocalization theory, *grobalization* is seen to encompass the following:

- The world is growing increasingly less diverse.
- Individuals and groups have relatively little ability to adapt, innovate and manoeuvre within a grobalized world. Larger forces and structures tend to

overwhelm the ability of individuals and groups to create themselves and their worlds.

- Grobalization tends to overpower the local level and limits its ability to act and react, let alone act back on the grobal level.
- Commodities and the media are the key forces and areas of cultural change, and they are seen as largely determining the self and groups throughout the grobalized areas of the world.

In relation to this last point, *glocalization theory* (Robertson, 1994; 2001) also sees commodities and the media as key forces in cultural change in the late twentieth and early twenty-first centuries. However, in stark contrast to the grobalization position, these forces are considered to be providing material for use in individual and group creation throughout the glocalized areas of the world.

The grobalization theory of late modernity is quite in line with the analyses and descriptions carried out by the Nordic ecophilosophers some thirty years ago, in particular as it has been elaborated in the works of Kvaløy. And of course grobalization forces expanding and imposing on the local, generate severe ecological imbalances and environmental problems, locally as well as globally. This is not included in the Ritzer theory. To him nature is largely non-existent; in this he of course follows a long tradition in sociology, carrying on the inheritance both from Durkheim and Weber.

Globalization of nothing

Ritzer (2004a) also launches the thesis of globalization as production of *nothing* and *nothingness*. *Grobalization* leads to an increasing dominance of nothing in the form of non-places, non-things, non-people and non-service, all at the expense of something on a nothing-something continuum. All of us do of course 'meet' non-people every day, through Internet communication and when trying to talk to answering machines on the phone. Non-places of late modernity are for example major highway intersections, highway motels and international airports. Here we of course also often meet non-things, non-people and non-service. The French anthropologist Marc Augé (1995) defines a *non-place* as a space which cannot be defined as relational, or historical, or concerned with identity. To describe such places he uses terms as solitary, fleeting, similitude, anonymity, lack of history, etc. Ritzer (2004a, p. 20) connects the following main characteristics to the nothing-something continuum:

Something	**Nothing**
Unique	Generic/Interchangeable
Local Geographic Ties	Lack of Local Ties
Specific to the Times	Time-less
Humanized	Dehumanized
Enchanted	Disenchanted

According to Manuel Castells (1996) we are in the so-called network society moving from a world characterized by 'spaces of places' to one dominated by 'spaces of flows'. There is of course a high correlation between spaces of places and something on the one hand, and *spaces of flows and nothing* on the other (Ritzer, 2004, p. 41).

In their ecophilosophy both Zapffe and Kvaløy put emphasis on the production of nothing, non-places and non-people. But in late modern sociology, terms like spaces of flows, nothingness and non-places often convey the post-modernistic illusion that this is totally *foot-loose* in relation to nature; that these are processes that also represent nothing in relation to nature. Quite contrary to that, there is nothing that can imply ecological neutrality. Production of nothing definitely leads to something in nature, more severe forms than before. The global flows are fortifying processes of ecological crisis. Non-places are integrated in particularly environmentally harmful societal processes, as *hypertourism, hyperconsumption* and *hypermobility* in the form of extreme automobility and aeromobility, as I have termed it.

Enchanting a disenchanted world, cathedrals of consumption and ecological crisis

George Ritzer (1998; 1999; 2004b) launched other concepts too. Consumption in late modernity takes increasingly the form of *hyperconsumption*. To serve this form of consumption, new *means of consumption* are developed, and globalized, and come to play important roles in the grobalization. They are becoming much more dominant than the means of production, in the Marxist connotation. According to the Ritzer analysis the new means of consumption should be seen as *cathedrals of consumption*. They have an enchanted, sometimes even sacred, religious character for many people. Travel to them often takes the form of *pilgrimages*. Several analysts contend that shopping malls, for instance, are more than commercial and financial centres; they have much in common with the religious centres of traditional civilizations (Ritzer, 1999, p. 8).

The main cathedrals of consumption are: super malls, shopping centres, casinos, cruise ships, amusement parks, international airports, major railway stations, football and athletic stadiums, but also many universities, museums, restaurants (*eatertainment*), Internet, and to really close the circle: even new mega churches, are going through a similar development. In order to attract ever-larger numbers of consumers, such cathedrals of consumption need to provide, or at least appear to provide, increasingly magical, fantastic and enchanted settings. One type of enchantment is making the shopping buildings ever larger. Not only malls, but now mega malls.

This week (in August), by the way, I heard on the news that the London megadepartment store Harrods had opened their Christmas sales. And as a commentary to what he termed a cynical (*sic*!) critique, their communication director emphasized that these sales only covered a minor part of the store. Not until November would the whole place be geared up for Christmas sales. A representative of the competitor store Selfridges says: 'Christmas is coming earlier each year. I can foresee a time when we offer a Christmas collection throughout the year.' Really a most logical conclusion. The post-modernist Baudrillard (2001) may be right after all: *Disneyland is there to hide the truth that it is the real country, all of the real America, which is Disneyland.* However, we must remember; all the carnivals, all of Disneyland, only take place on the surface. Beneath the surface there are the globally, totally disenchanting, standardized and engulfing processes of grobalization.

In their ecophilosophy both Zapffe and Kvaløy emphasize the importance of the *Disneyland effect* as a means to compensating the human tragedies of the industrial growth society; this is their description of the process of continuously enchanting a disenchanted world. But Ritzer extends this; there is a close relation between hyperconsumption and enchantment. The new cathedrals of consumption play crucial roles both as places for enchantment and as arenas for hyperconsumption. We know very well from current development trends in our affluent societies that shopping is not just fun, it is one of the most pleasurable indulgences around. But its effects on the environment are no fun at all. The cathedrals are closely knit to the particularly troublesome systems of hypermobility; the auto- and aeromobility. And all the commodities are produced from something, and bring something – rather problematic – back to nature during production, consumption and disposal.

Conference tourism and life in corridors

(Trying to use humor in line with a long tradition in ecophilosophy, even though human life is tragic, and there is an ecological tragedy) I have referred to the Kvaløy term of human life in the industrial growth society as life in a pipe system. In my own analyses of *conference tourism* I have launched the concept *life in corridors* to catch some major aspects of the lives of academics in a world of international conferences. As in the historical term the *Polish corridor* these are channels established through foreign territories to achieve highly efficient communication. But there is only communication along the corridor; it is closed to both sides. There is no communication to the territories they pass through.

This is the very condition for efficiency. And of course they have the asset of being collective routes for many people. However, they not only lack communication to the sides; all following the corridor are doomed to come to the same corridor ends, the definite destinations as was the case with the Polish corridor. In the life of a conference tourist there are corridors everywhere. Along the highways – or high-speed railways – to the international airports for departure. For the luggage on conveyor belts. Following all the others to the departure gates. Yes, this is exactly what it is called: *gates.* There is no airport term as important as this. Then, going through the gates for embarking. Following the others to your seat. Look along, and sometimes walk through, the very narrow airplane corridor. The airplane itself follows an air corridor, be it in the troposphere or the stratosphere. When landing, everything is repeated in reverse. Seemingly incessantly forwards and backwards, just as we have learnt in Newtonian physics. And you end up in an international hotel after having been carried along new corridors from the international airport you arrived at. But everything looks the same wherever you are in the world. It is really disenchanted. The airports are alike. The airport landscapes with their connecting transport corridors look the same. The planes are the same. The international hotels are all alike. Even the language is the same; the globalized *Anglo-American*, the *corridor language* for the whole world. It is a standardized language of nothingness.

This same language is also the *conference language* itself; also in the conferences where researchers gather to present their latest findings within the language turn in discourse analysis. There are not many differences between the various international conferences; though there are always some aspects of pleasant enchantment included to augment competition with other conferences. But in the disenchanted conference settings there are always too many paper presentations, too many sessions, and too little time for discussions. And as there are conference hoppers, there are professional session hoppers, only there to listen for a short time before running to another session. And of course the prime purpose is presentations, or perhaps rather to be *re-presented*; not the discussions per se nor deliberations, as the post-modern term would be. Anyhow, if you have *jet lag*, the human brain is so wisely constructed that you probably have lost your short-term memory; a loss that lasts just as long as the conference itself, two to three days. Conference tourism is globalized as most other major forms of tourism; it is part of the globalization of academia, and it serves to turn academicians into players in the processes of grobalization. Conference tourism is a global industry where competition on a global market is an important factor. They may be producing nothing. But they definitely lead to something, in the basic ecological systems.

The corridors are closed to the sides in every respect, also regarding the possibilities to experience effects on ecology. A *tunnel view* on reality is a necessary implication of being a busy corridor traveller. But in an age of *ecological enlightenment* we – as academicians – are all very well informed about the severity of these effects. Or are we? There is no other form of mobility bringing with it a similar seriousness of ecological problems. And there is no other form of grobalization as global tourism with similar contributions to the ecological crisis. And conference tourism is only a part of global tourism. In no other field are there larger ecological load differneces between the highly mobile global elite and the vast immobile majority of the world population.

Notes

1. The collected writings of *Peter Wessel Zapffe* has been published (only in Norwegian) by Pax Forlag; see Zapffe (1996; 1997a; 1997b; 1998a; 1998b; 1998c; 1999a; 1999b; 2002). I know of only one text translated to English, published in Reed and Rothenberg (1987).
2. *Biosophy* on the international scene is defined as a humanist movement most heavily influenced by Spinoza. It is considered to be 'the science and art of intelligent living based on the awareness and practice of spiritual values, ethical-social principles and character qualities essential to individual freedom and social harmony'. Among the objectives of the current international biosophy programme are:

- to create biosophical groups for character and peace education where the individual has opportunities of self-improvement;
- to replace the blind acceptance of theological beliefs, superstitions and dogmas with the cultivation of spiritual intelligence as the basis for the Religion of Freedom and Friendship;
- to create a world-fellowship of peace-loving human beings who have overcome religious, national, racial and social prejudices who can work creatively for the growth of democracy, spiritual personal growth and world peace;
- to work for the synthesis of religion, philosophy, science, education and art;
- to cooperate in the advancement of humanity from civilization to the age of soulization.

3. Regarding the many quality levels in complexity, Kvaløy (1973a, p. 150) emphasizes that intuition, sensitivity, love, direct integration of body and mind in rhythmic movement with nature and fellow beings, aesthetic judgment as a force in practical matters, a person's or group's attachment to a specific home environment, the inner strength and self-reliance they are given through this attachment, etc., all are examples of human capacities belonging to these other levels. 'Nature and its child – the human being – are continuously working on such levels; – areas of function and relationships that do not yield to readymade measurable analysis; and neither the ecosystem nor the human being can do without these modes of functioning this side of disaster.'
4. The Swedish professor in the history of ideas – Sverker Sörlin (1991) – has termed our current processes in nature conservation and environmental politics for something like a 'Du contrat nature'; and he compares this with early social processes connected to the Rousseau works on 'Du contrat social'.
5. Connected to a Nordic research project I participated in (1989–91) I have earlier published about this topic – the new environmental organizations (Høyer, 1991; Høyer and Selstad, 1991). Similar viewpoints as expressed here have been published by the Norwegian sociologist Frode Gundersen (1989). My later sources – post 1990 – for these viewpoints are mainly participatory observation at conferences and through lectures and ensuing discussions with members of the various environmental organizations.

References

Adam, B. (1996): Re-Vision: The Centrality of Time for an Ecological Social Science Perspective. In: Lash, S. *et al.* (eds): *Risk, Environment & Modernity. Towards a New Ecology.* London: Sage Publ.

Andersen, G. (ed., 1990): *Momenter til en dypere naturvernforståelse.* Østlandske Naturvernforening. Oslo: Vett & Viten Publ.

Augé, M. (1995): *Non-places: Introduction to an Anthropology of Supermodernity.* London: Verso.

Baudrillard, J. (2001): Simulacra and Simulations. In: Poster, M. (ed.): *Jean Baudrillard. Selected Writings.* Cambridge: Polity Press.

Beck, U. (1986): *Risikogesellschaft: Auf dem Weg in eine andere Moderne.* Frankfurt am Main: Suhrkamp Verlag.

Beck, U. (1995a): *Ecological Enlightenment.* New Jersey: Humanities Press.

Beck, U. (1995b): *Ecological Politics in an Age of Risk.* Cambridge: Polity Press.

Castells, M. (1996): *The Information Age, 1. The Rise of the Network Society.* Oxford: Blackwell.

Georgescu-Roegen, N. (1971): *The Entropy Law and the Economic Process.* Cambridge: Harvard University Press.

Georgescu-Roegen, N. (1972): Economics and Entropy. *The Ecologist.* July 1972. London.

Georgescu-Roegen, N. (1975a): Energy and Economic Myths. I. *The Ecologist*. Vol. 5, no 5. London.

Georgescu-Roegen, N. (1975b): Energy and Economic Myths. II. *The Ecologist*. Vol. 5, no 7. London.

Gundersen, F. (1990): Den moderne miljøbevegelsen i Norge. In: Andersen (1990).

Hägerstrand, T. (1981): Space and Time in Geography. Essays Dedicated to Torsten Hägerstrand. *Lund Studies in Geography no. 48*. Lund: C W K Gleerup.

Hägerstrand, T. (1987): Torsten Hägerstrand Laudatio. *GeoJournal*, 14.4. Boston: D. Reidel Publ. Company

Hägerstrand, T. (1988): Spatial Diffusion. *Scientific Geography Series*, 10. Dedicated to Torsten Hägerstrand. Newbury Park, California: Sage Publ.

Hägerstrand, T. (1991) *Om Tidens Vidd och Tingens Ordning. Texter av Torsten Hägerstrand*. ed. Carlestam, G. and Sollbe, B. Rapport T21: 1991. Stockholm: Statens Råd för Byggnadsforskning.

Hägerstrand, T. (1993): *Mobilitet*. Teknikdalen Borlänge, Sverige: Johan Gottlieb Akademien.

Hellesnes, J. (1996): Introduksjon. In: von Wright (1996).

Hofseth, P. and Vinje, A. (eds, 1975): *Økologi – Økofilosofi*. Oslo: Gyldendal.

Høyer, K.G. (1979): *Olje, energi og planlegging*. Oslo: Universitetsforlaget.

Høyer, K.G. (1991): Regionalpolitikkens økopolitiske grunnlag. In: Elling, B. *et al.*: *Regionalpolitikkens økopolitiske grunnlag i Norden*. NordREFO rapport 1991: 5. København: NordREFO.

Høyer, K.G. and Aall, C. (2005): Sustainable Mobility and Sustainable Tourism. In: Hall, M.C. and Higham, J. (eds): *Tourism, Recreation and Climate Change*. Clevedon: Channel View.

Høyer, K.G. and Selstad, T. (1991): *Regionalpolitikkens økologiske grunnlag*. Vestlandsforsking/Østlandsforskning/NordREFO, VF Rapport 19/91. Sogndal: Vestlandsforsking

Kvaløy, S. (1971): *Økofilosofisk Fragment I*. Oslo: Universitetet i Oslo, Filosofisk Institutt.

Kvaløy, S. (1973a): *Økofilosofisk Fragment IV*. Trondheim: Tapir Forlag.

Kvaløy, S. (1973b): Økokrise, Natur og Menneske. *Nordisk Forum, Hæfte 2*, 1973. København: Munksgaard.

Kvaløy, S. (1975): Økokrise, natur og menneske. In: Hofseth and Vinje (1975).

Kvaløy, S. (1979): Fra Industrivekst til Økopolitikk. In: Berntsen, B. (eds): *Norsk Økopolitikk*. Oslo: Universitetsforlaget

Kvaløy, S. (1985a): *Grønn pepper i turbinene* – til ære for Sigmund Kvaløy Setereng. 16 bidrag om miljøkrisen og veiene ut av den. Redigert av S.E. Skønberg. Oslo: Universitetsforlaget

Kvaløy, S. (1985b): Mangfold og tid. Hvordan en bryter pyramidens herredømme. In: Kvaløy (1985a).

Kvaløy, S. (1990): Økofilosofi versus New Age – friluftsliv kontra kvantesprang. In: Andersen (1990).

Kvaløy, S. (1991a): Systemkrise og krisesystem. In: Album, G. *et al.*: *SuperMarked eller Felles Framtid. EF, Økokrisen og Norges valg*. Oslo: Cappelen.

Kvaløy, S. (1991b): Økokrisefilosofi – glimt fra det norske økofilosofiske forsøket. In: Gjerdåker, S. *et al.* (eds): *Den Uoverstigelige Grense*. Tanke og handling i miljøkampen. Oslo: Cappelen.

Kvaløy, S. (1992): Identitet og meningsfullt arbeid. In: Stenseth, N.C. and Hertzberg, K. (eds): *Ikke bare si det, men gjør det!* Om bærekraftig utvikling. Oslo: Universitetsforlaget.

Kvaløy, S. (2000): Gaia versus servoglobe. In: Dybvig, M. *et al.* (eds): *I et filosofisk terreng*. Festskrift til Sverre Sløgedal. Trondheim: NTNU, Filosofisk Institutts Publikasjonsserie nr. 34.

Kvaløy, S. (2002): To Økofilosofier i Norge; Deres begynnelse og en del til. In: Næss (2002).

Kvaløy, S. (2004): *Mangfold og Tid*. Økofilosofisk fragment nr. 7. Trondheim: Tapir akademisk forlag.

Næss, A. (1972): *Økologi og Filosofi*. Et Økosofisk Arbeidsutkast. Preliminær utgave. Oslo: Universitetsforlaget.

Næss, A. (1974): *Økologi, Samfunn og Livsstil*. Oslo: Universitetsforlaget.

Næss, A. (1975a): Økosofi T. In: Hofseth and Vinje, A. (1975).

Næss, A. (1975b): Vitenskapsmannen – en anonym profet? In: Hofseth and Vinje (1975).

Næss, A. (1985): Mangfold og tid. En kommentar. In: Kvaløy (1985a).

Næss, A. (1989): *Ecology, Community and Lifestyle*. Translated and edited by David Rothenberg. Cambridge: Cambridge University Press.

Næss, A. (1993a): The Deep Ecological Movement: Some Philosophical Aspects. In: Armstrong, S.J. and Botzler, R.G. (eds): *Environmental Ethics. Divergence and Convergence*. New York: McGraw-Hill, Inc.

Næss, A. (1993b): Beautiful Action. Its Function in the Ecological Crisis. *Environmental Values*, 2.1.

Næss, A. (1997): The Shallow and the Deep, Long-range Ecology Movement: A Summary. In: Fjelland, R. *et al.* (eds): *Philosophy beyond Borders: An Anthology of Norwegian Philosophy*. Bergen: SVT Press, University of Bergen.

Næss, A. (1999): Articles in: Witoszek, N. and Andrew Brennan (eds): *Philosophical Dialogues: Arne Næss and the Progress of Ecophilosophy*. Boulder, USA: Rowman & Littlefield.

Næss, A. (2002): Arne Næss. Festskrift til 90-års dagen 27.01.2002. *Norsk filosofisk tidsskrift*, nr. 1–2, 2002, årgang 37.

Reed, P. and Rothenberg, D. (eds) (1987): *Wisdom and the Open Air: Selections from Norwegian Ecophilosophy*. Oslo: Rådet for natur- og miljøfag, Universitetet i Oslo.

Ritzer, G. (1998): *The McDonaldization Thesis*. London: Sage Publ.

Ritzer, G. (1999): *Enchanting a Disenchanted World: Revolutionizing the Means of Consumption*. London: Pine Forge Press.

Ritzer, G. (2003): Rethinking Globalization: Glocalization/Grobalization and Something/Nothing. *Sociological Theory* 21 (3): 193–209.

Ritzer, G. (2004a): *The Globalization of Nothing*. London: Sage Publ.

Ritzer, G. (2004b): *The McDonaldization of Society*. Revised New Century Edition. London: Pine Forge Press.

Robertson, R. (1994): Globalisation or Glocalisation? *Journal of International Communication*, 1, pp. 33–52.

Robertson, R. (2001): Globalization Theory 2000+: Major Problematics. In: Ritzer, G. and Smart, B. (eds): *Handbook of Social Theory*. London: Sage Publ.

Sörlin, S. (1991): *Naturkontraktet*. Stockholm: Carlssons.

Urry, J. (2003): *Global Complexity*. Cambridge: Polity Press.

Von Wright, G.H. (1986): *Vetenskapen och förnuftet*. Månpocket. Stockholm: Bonnier Fakta Förlag.

Von Wright, G.H. (1993): Vitskapen, mennesket og miljøet. *Syn og Segn, hefte 4*/1993.

Von Wright, G.H. (1994): *Myten om fremskrittet*. Oslo: Cappelen.

Von Wright, G.H. (1996): *Å forstå si samtid. Essay*. Oslo: Det Norske Samlaget.

Zapffe, P.W. (1969): Farvel Norge. Trykket. In: Zapffe (1997b).

Zapffe, P.W. (1975): Et biosofisk perspektiv. In: Hofseth and Vinje (1975).

Zapffe, P.W. (1989): Ambolten Hvorpå Selv Gudene Hamrer Forgjeves. Sigmund Kvaløy om Zapffe. Norsk Filosofisk Tidsskrift. *NFT 24 (1989) 225–241.*

Zapffe, P.W. (1996): *Om det tragiske*. Oslo: Pax Forlag.

Zapffe, P.W. (1997a): *Kulturelt nødverge. Etterlatte skrifter.* Oslo: Pax Forlag.
Zapffe, P.W. (1997b): *Barske glæder.* Utvalg ved Sigmund Kvaløy. Oslo: Pax Forlag.
Zapffe, P.W. (1998a): *Den logiske sandkasse.* Oslo: Pax Forlag.
Zapffe, P.W. (1998b): *Spøk og alvor.* Oslo: Pax Forlag.
Zapffe, P.W. (1998c): *Skuespill. Den fortapte sønn og Hos Doktor Wangel.* Oslo: Pax Forlag.
Zapffe, P.W. (1999a): *Indføring i Litterær Dramaturgi.* Oslo: Pax Forlag.
Zapffe, P.W. (1999b): *Hvordan jeg blev så flink.* Oslo: Pax Forlag.
Zapffe, P.W. (2002): *Vett og Uvett. Stubber fra Troms og Nordland.* Oslo: Pax Forlag.

5

A BIOSOPHICAL PERSPECTIVE

Humans as a tragic species

Peter Wessel Zapffe (Introduction by Sigmund Kvaløy)

Peter Wessel Zapffe was probably – with his *biosofy* – the leading philosopher in Norway to have a committed and clearly elaborated eco-philosophy of his own. The ideas that the ecological crisis of Western civilization have given such urgency are fundamental in his work from as early as the 1920s – long before he began using the word 'ecology'. It is clearly demonstrated in a piece written (in poetical mode) in 1929 for the publication *Tromsø Stiftstidende*.[1] His thinking, even at this early stage, tallies with modern system ecology, and was in all probability developed independently of it. As in system ecology, Zapffe grounds the dynamics and existential conditions of life in reflections on the so-called law of entropy – the theory of the inexorable tendency in the universe to even out energy differences and replace order with chaos – that is, to increase the amount of entropy.

All humanity, and every possibility of realizing values, depend entirely on mankind's ability to resist this tendency – precisely what life on the planet in general has succeeded in doing for three billion years. Progress in light of this existentially unique counter force – the development of an ever strengthening structured diversity, depends in turn on the unremitting decomposition suggested by the entropic tendency. Growth builds on decay, death makes life possible. Only when entropy has done its work will the sustained, irreversible change in our established states enable further diversification by exploiting as yet unused reproductive abilities and potentials inherent to organic matter.[2]

This conception of the singularity of life might perhaps be pleasing in an aesthetic or religious sense, but the individual person will not necessarily be persuaded as long as it maintains a grip on the affective consciousness which itself is inseparably bound up with the same relentless stream. Duration, understood as ecological balance, is not the same as the static states which ordinarily and unthinkingly provide us with a sense of confidence. On the contrary, the biological process's achievement of 'equilibrium through change'[3] gives us certainty about one thing only – that growth is followed by decay and that new forms of life must replace the old. The achievements of mankind must fall, and the noise will be all the more deafening the mightier the constructed edifice.

The ecological crisis is a pertinent example, and explains Zappfe's horror of the destruction of the natural environment by industrialized societies and unconscious compulsion in adherents of economic-growth to succeed. It gave his justification of conservation a wider applicability and depth, things which only in recent years have gained a wider currency.

More unusually for Zappfe is how he expands the scope of systemic/entropic thinking to people's mental life – as the excerpt from *Indføring i Litterær Dramaturgi* [*Introduction to Literary Dramaturgy*] below will make clear: the more meaning and justice people want, the more people must be prepared to face disappointment. Nature is not designed to meet these demands and their ilk. At the same time, they are precisely what define humanity's singularity. If mankind relents on these demands, it relents on its humanity. The basic principles of biological dynamics – what differentiates life from everything else in the universe – see humanity then as a chance offshoot in the biological undergrowth. It is through this that humanity fulfils the terms of which it means to be a 'tragic' being, as Zappfe defined it in his doctoral dissertation from 1941.

Excerpt from *Introduction to Literary Dramaturgy*, Oslo 1961

'Biosofy' means a philosophy, a theoretical conception of existence, oriented towards biology and its related disciplines (palaeontology etc.), and therefore deals with every 'problem' in a completely empirical fashion. The biosophical conception of a phenomenon differs therefore significantly from a metaphysical (extending to spiritualistic, materialistic, vitalistic, finalistic, religious, etc.) as much as a speculatively aesthetic holism – to which it remains indifferent, and not necessarily polemical. If, then, these holistic pictures are such that the unknown forces can be taken as manifesting themselves *through* the 'laws of nature' and the 'biological life', they rely then on the same observational data and the difference lies only in their interpretation. Not so if impulses are included, which for conscious 'teleological' reasons have butted against and may break through at any moment and change the expected course of events in light of common experience. The difference lies in the deepest preconditions, and insofar as no common platform exists, no compromise is possible. A 'belief' located in extension to experience, is of interest only as a curiosity, and a belief which runs counter to experience (God ensures all earthly requirements are met, even without people having to work), is of interest only as a symptom. Biosophically relevant reports can be found in V. Brøndsted, *Biologi og menneskeforståelse* [*Biology and Perceptions of Humanity*], and J.v. Uexküll, *Biologische Weltansschauung* [*Biological Worldviews*].

Given the question of dramatic *impact*, we trespass necessarily into *psychology's* domain, while the dramatic *material* has to be quickened where Life struggles to survive and advance in a more or less favourable environment. All biosophically inclined scientists will instinctively be drawn to disciplines whose subject matter is precisely this struggle for survival.

Here, as so often in other contexts, we come across the 'purely accidental', the pedagogically speaking most accessible material in its *primitive* expression of life. Biology presumes a certain basic kinship between all living organisms with regard to endowment and conditions, differences apart. A biosophical aspect was, by the way, prefigured by the prophet *Jesus*, who in the fable of the seed which fell in different places, in less than a nutshell, in a husk of corn, offered a general formula for life and its conditions. (The properties of the corn are assumed to be identical, and no Jehovah intervened.) Eighteen hundred years later, but before Darwin, Schelling-Steffens-Wergeland saw in [the poem 'The Creation – Man and Messiah'] 'Ormens snelle Bugten' ['the swiftly slithering snake'] the angel struggling to stretch its wings – a thought confirmed by evolutionary theoreticians in respect in the path from amoeba to genius. The basic similarities in properties and conditions can be observed by anyone, independent of theory.

The traditional 'individual–environment' dualism is not, however, as effective as it was before Jacob von Uexküll's reasoned thesis in *Umwelt und Innenwelt der Tiere* (Berlin 1921) [*The Environment and Subjective World of Animals*] that the 'environment' as perceived by the individual species, indeed the individual organism, is partly and possibly completely determined by the sensory properties of living organisms, possibly in association with a subjective rendering of the impressions. These abilities and properties are the only means by which the properties of 'environment' can be identified – or more precisely: by which an X can be determined, and subsequently named or treated as 'environment'. As far as ordinary experience is concerned, this ability found a colloquial expression in the phrase: it depends on who's looking. In recent years, the radio apparatus is frequently used as a visual model: it only catches the waves to which it is tuned, the rest is silence, it 'doesn't exist' for the apparatus. None of these waves exist as immediate sensory impressions in people of course, only through 'knowledge' and the apparatus as a medium, as a supplement and, therefore, a modification of our sensory equipment. The dog lives in a world of sounds and smells; we know them only by inference etc. We can approach Uexküll's essential idea in light of Berkeley's thesis, 'esse et percipi' – and we can even trace it back to Protagoras' 'homo-mesura doctrine' (A. *Næss*, *Filosofiens historie*, Oslo 1967, pp. 74ff.). In various guises, the subject of 'me and the world around me' is present in nearly all philosophy. As long as one 'thinks biosophically', the architecture of philosophical formulations is of secondary interest. The tenability of the thesis 'esse et percipi' (to be is to be perceived by a sentient being) must be judged (after specifying the idea) from a point outside 'percipi', from which an 'esse' can be observed with 'percipi'. But we would then have refuted the thesis before we had an opportunity to compare the discovered 'esse' with its 'perceptum' and settle the identity question. Following Uexküll's principle, the aphorism could be rephrased to greater advantage thus: 'Anything but perceptum is beyond the scope of biology's remit.' But at the point of convergence (possibly cumulation) of several perceptions, insofar as a Berkeley or an Uexküll disciple assumes his perception to include the animal's, therein lies at least an initial complication ('perceptum est percipi').

In practical life, confusing a perception with an assumed 'esse' can have very serious repercussions. However, if a different individual entertains a different perception, it is not, biosophically, immaterial unless the perceived object acquires for the person a different interest-related relevance.

A living being is characterized in biosophical terms by its *interests*, in relation to which the sensory apparatus with appurtenant intellect, memory, imagination and emotions is secondary.

The word 'interest' is used here as a generic term for several more or less synonymous terms: need, drive, desire, wish, 'will', etc.; it is not for the moment to insist on distinctions. Interest manifests itself when something becomes 'urgent', 'of the essence', when it is important, weighty. The individual will, more or less intensely, coherently or divided between inclinations, fancies, seek to 'achieve something' or 'avoid something'. It 'likes' or 'detests', it says Yes or No to respectively sustaining or changing a state. (We could possibly organize causally disposing interests in 'binary dichotomies': Yes to contact, No to isolation, etc., but the experiment is not particularly compelling for the moment.)

Individuals acquire, in other words, biosophical relevance insofar as they act as *interest-carriers*. The compound noun, despite its central place in the train of thought per se, should not require further explication. As an expression it has several advantages. It covers the individual as biosophical entity and does not require a stand on disputed dualisms such as 'spirit and matter', 'body and soul', 'materialism and spiritualism', 'law of psychic causes or free will', etc. Distinctions of this variety become biosophically relevant only if they are substantively essential to a person's well-being. (Cf. *Uexküll*, Biologische Weltansschauung, Munich 1913, p. 143 *passim*, V. Brøndsted, Vitenskap og livssyn, p. 60.)

For *philosophy*, the germane questions are the subject of ongoing work, whose completion is an end in itself or a means to other ends (progress and the like). The *dramaturge* or *dramatist* does not encounter philosophers very often, though occasionally, a dramatically inclined poet may make 'causal relations' the subject of a particular discourse. In the most frequently occurring 'dramatic material' (here – material used in or probably suitable as a drama) he deals with people whose interests do not include solving theoretical problems, such as those precipitated by the aforementioned dualisms. He may perceive relations such as 'body and soul' as an aid in his interpretation of concrete, urgent situations. Questions concerning causal law and mankind's 'free will' are a particular case – partly because they have been a traditional concern throughout the history of philosophy (metaphysics, ethics, criminal law) and partly because every single human being has to deal with, more or less daily, events of fundamental importance and the perturbing, emotive practical problems they throw up. We face a choice, for example, to pursue a course of action with strong reasons for and against, possibly with fateful consequences for both sides. Or we might be asked to assess another person's actions. Experience tells us that our actions are affected by 'dependency factors', 'adequacies', 'compos mentis', probability, 'functional relations', etc. We are all familiar with the ethical struggle that comes with unavoidable decisions, but we don't bother our minds with what someone might argue in a discussion of the 'free will' – for instance the idea that a 'willed choice without reasons' or 'independent of motives' escapes conceptualization and seems absurd, or the opposite, that it is 'degrading and outrageous for a free person' to subject himself to the dictate of the causal masses. In fraught situations, theoretical speculations are a luxury prohibited by lack of time. To hesitate becomes to act. The dramatist is dealing here with interest-carriers, whose interest in the questions' theoretical resolution occurs sporadically and in line with other interests.

By 'interest-carrier' we are thinking essentially about individuals who are aware or conscious of their interests. We could also say definitionally that interest does not occur before the awareness of a tendency, etc., exerts pressure on the decision-making apparatus, which will then proceed to realize the interest by whatever means are available. In lower animals and plants, which do not in our opinion have the capacity of choice, it would be more sensible to speak of 'conditions of life'.

Every interest-carrier exists in an *environment*, whose relations to his current interest status differ and fluctuate. The environment may *harmonize* with the interest (contact, sympathy, love, solidarity, etc.) – the normal state in parents as qualities of their offspring's environment. It may be *indifferent* (earth, water, air, certain animals and people who 'don't care' about him. They can harm or promote his well-being, though they are not, as far as he can judge, a cause of action). (Photography of 'indifferent environment' in Windisch, Fargefotografi, Oslo 1954, p. 105.) And it can be *hostile*, evil, 'diabolical', which it is when its interests coincide with the abuse of the mentioned individual's interests (sadists, etc.). Wild animals compelled by hunger to attack, or leaders who reluctantly take disciplinary measures can be placed in an intermediate class.

By destiny we understand here the relationship between the interest's present state, and an envisaged, desired, dreamed, hypothetical, ideal state. In the unavoidability of some destiny or other, to the individual's delight, indifference or abhorrence, there is a community of all interest-carriers, from the rats near the rubbish bins, to the wrestling of the genius with the latest answer. In biosophical light, beings, things and factors are considered *sources of destiny*. Insofar as they stop being that, they approach the periphery of the biosophical field of vision. Hence, for instance, a 'global plan' in which the conscious person does not participate apart from providing impersonal material.

Only interest-carriers have a destiny. No one experiences 'differences in contentment' in an interest-less world, whatever happens to them – for instance that entities defined by mankind such as 'moon', 'ice block', 'potato' merge or separate, are crushed or scattered, burn up or freeze over. Many such entities are not even credited with a 'history' in the sense of 'identity with variations'.

Conversely, the word 'potential' alludes to a community uniting the entire known world. The term 'potential' can signify, for example, 'accumulation of energy', gravitational potential (a rock in a high place), pressure potential (in a pressure cooker), heat potential (in a heat storage cooker), electric potential (in a charged pole). All potentials are similar in that as long as they meet no resistance they will 'leak', be 'released', etc., until the higher potential equals the environmental potential. On the other hand, nature does not allow a potential to grow or accumulate 'by itself' at the cost to the environment. The laws say Yes to equalization, and No to accumulation. To compress energy in violation of nature's will, supplied units of power must be supplied and consumed, and spent relative to the purpose (pressure pumps).

Living beings (here interest-carriers) have their own purely physical energy requirements which are met by ingesting nutrients, which in turn are sustained, at the end of the day, by solar radiation. But also in the figurative sense, a 'life potential' can be imagined in relation to dead or less differentiated nature – though a picture whose content and efficiency vary in the extreme. Parallels with physical potentials should be handled very circumspectly, but that is no reason to avoid them altogether. Fluctuations in a person's 'psychic energy' often evade any sort of study, but a strong emotion and its translation into words or deeds feels immediately like the charging or discharging of something, and corresponding verbal metaphors are widely used in everyday speech.

The balance between a high and a low potential is something physicists call 'charge' – cf. the drawn bow. In a charged relationship, three factors will be present: the charged energy; the surrounding field; and the factor preventing discharge. *Interests* can be considered highly potential relative to a world (not where interests are automatically satisfied, as was the case for the absolute monarchs, but) where the supply of energy cannot be interrupted (possible isolation), and even more to realize themselves. There is a risk attached to many strong, conspicuous interests that things will not go in the interest's favour. The 'lower potential' of the environment, possibly the opposite form of energy (positive and negative charges), acts like gravity, a pulling force. It is far more demanding to build 'values' (interest-serving states) up than to destroy them (the city and the bomb). Diseases invade the body in 'yards' but must be evicted 'inch by inch'. A long-standing friendship runs aground on a trifle. If I am on the third floor, I can 'in the twinkling of an eye' let myself down and be afflicted by the greatest remorse and suffering. But a reverse 'twinkling' will not bring me up to the fifth floor and a corresponding sense of ecstatic elation. The 'values' need constant supplies of energy from the outside (for example financial profits) while 'nature takes charge of' waste matter.

People with strong, rare and vulnerable interests tend to surround themselves with people of a similar constitution (surroundings with the same potential) because an indifferent or an otherwise minded environment will exhaust, undermine, the realization of the interest (inventors, artists and the like).

The study of these things is called *ecological* biology or bionomics. If we let the words cover all 'struggles of interest' from which the 'survival of the fittest and the luckiest' is the outcome, the material-dramatic aspect (the 'objective dramatic aspect') can be said to belong to the ecology of human life.

And in what follows our only concern will be with human life, even though some of the illustrations are derived from the animal kingdom.

We shall call the total sum of interests of an individual or group at a given point in time the individual's or the group's *interest front*. A study of the conditions for life in the lower, and of interests in the higher animal kingdom would be an arresting supplement to the doctrine of descent and evolution. But insofar as the dramatist in the event only to deals with people, we shall not retrace our steps further than to primitive man.

Roughly, then, the interest front consists of concerns such as these: health and physical integrity, freedom, good relations to members of the tribe, air, food, heat, sex and offspring, 'pleasant atmosphere' for no good reason, things which taste, smell nice, sound lovely, look nice, etc. Each of these concerns exist in a given environment whose conditions differ and vary, beneficially or detrimentally, and their total status determines the individual's well-being, happiness and sense of fulfilment.

When the individual is concerned with fulfilling a (possibly threatened) interest (his desire, attention and capacity join forces in facilitating fulfilment) – he tries, for instance, to provide food for himself and his loved ones by hunting and fishing – the situation can be described with the help of the following factors:

1. The *interest or need* in question: need of food due to hunger which is a state he wants to alleviate.
2. The *object* or *thing*, here wild animals, the death of which signals the satisfaction of a need (the current *task*).
3. The *tools* available to the hunter (weapons, muscular energy, but also skills, practice, intelligence, cunning, swift in action).
4. *External conditions* (landscape, weather conditions, etc.).
5. The concomitant *risk*. This comprises not only risk of his failing with the expenditure of energy exacerbating his hunger and tiredness, but the animal, which itself is an interest-carrier, and whose interest differs from or positively the reverse of the hunter's, could, alone or aided by his fellow beasts, attack and destroy things of value which until then were intact.

As factor no. 6, we can include the hunter's *attitude* during and after the undertaking (joy of the hunt, foolhardiness, fear, cowardice, treachery against comrades, chance interpreted as skill, humble and honest, or strident, possibly a fraudulent record).

The *outcome* can be any one of three things:

a. The hunter is successful, he gets food for himself and his kin, he feels happy and proud and is congratulated by his tribesmen.
b. The situation is more or less as it was quo ante bellum, either with relief (the animal was overpowering) or disappointment (he should have brought it down).

c. The man is injured or killed, either without accomplishing anything or such that his efforts mean his family or tribe are saved (or a brave attempt has been made to save them) (the hero).

Of these engagement's 'component parts', nos 2, 3 and 4 (purpose [task], tools and external conditions) are variously mutually related on issues of balance, surplus or deficit (sufficiency, insufficiency, benevolence, malevolence. Cf. the author's 'Om det tragiske' ['On the Tragic'], Oslo 1941, paras. 11, 17, 47, 49, etc.) – a state of affairs liable to display surprising complications. A 'too much' can be as detrimental as a 'too little'. I might floor the hare by tossing a hand grenade (superior means), but there'll be nothing left for dinner. The polar bear is too large for me to hazard shooting it with bow and arrow. This is my salvation, because as it happens all the big bears around here are full of roundworms, etc.

This basic schema, culled from the primitive battle of interests, is still useful even after culture began to unfold with doubling and quadrupling of the population, inducing great changes technologically and psychologically.

1. *Needs*. All the primitive needs remain intact, but undergo increasing differentiation. Health alone occupies a wide front, exercise, bathing, dietary considerations, bacterial protection, mental hygiene, etc. Sex constitutes another broad, autotelic area, where erotic games unfold in richly shifting nuances, it is also made the subject of commerce in connection with theatre, films, illustrated magazines, gramophone records, cosmetics, etc. It is consciously associated with procreation and its control. The growth in the number of people, money economy and population concentration in cities facilitates the growing domination of the social aspect of the interest circle. Increasing numbers of concerns affect the individual's favourable relations with 'society', with employers and colleagues, family, judicial system, occupational groups and other organizations. The *ethical* area branches off and grows increasingly complex. Groups held together by some joint interest or other stand together or against other organizations, men and women, young and old, town and country, upper and lower class in different permutations of the word, confessions of faith, philosophical and scientific approaches, political parties, races and nations.

Technology along with more and more spare time and purchasing power permits ever wider sections of the population also to maintain an *autotelic* interest front – of varying cultural significance. Character development (differentiating and perfecting intelligence, feelings, imagination, taste, qualifications, skills and personality) are ends in themselves, in some cases the same can be said of science and art. And on top of it all, there is an increasingly clamorous demand for more and more opulence and entertainment, both 'higher' and 'lower' brow.

The *metaphysical* aspect of the interest front consisted originally of good relations with the weather, hunting and fertility gods, but also in an initially vague sense of wonder about death and the fate of the dead. One could venture to say that the biological instinct for self-preservation, autotelic urge to grow, and hunger for power, not to mention the metaphysical itch, are the interest fields that best depict the history of the Western cultural sphere over the past three thousand years. (Spontaneous *joie de vivre*, ethical idealism, etc. are secondary.) We could debate which of them had the most grotesque historic form; a thesis on the subject might give first place to the metaphysical sector. At base here is a vast desire for certainty on the central questions about the quiddity of the world, of life and death (their distinctive nature vis-à-vis humanity), their 'meaning' or 'meaninglessness' in light of *our* categories – usually gathered together in the question of the existence of a personal and omniscient god and his (the head god is always male) other attributes, in their relation to mankind's mortal and, possibly, post-mortal destiny.

Taken together these concerns all constitute in their infinitely varied manifestations a considerable burden on the mind of the civilized and culture-bearing person. After untold generations of intermixed hereditary traits they are ready to intervene in the new environment full of prospects. The most violent conflicts and contrasts are likely, not merely between people, but probably more often inside the individual mind. Individuals with richly constituted emotional lives, highly developed intelligence and global awareness can engage with concerns which for them are of vital importance, but which the primitive and powerful majority fails to comprehend and views therefore with loathing. These tensions can disperse within the collective mind at the instigation of all manner of social 'crises', for example when a new political ideas, perceptions of style, etc., supplant older ones.

2. The development of *objects* is similar. There is interaction at the interest front, not only because objects are hunted in nature, but are also manufactured in increasing quantities with the use of technology. In many areas, objects are surplus to demand, so advertisers are called in to create artificial needs. However, the prevailing principles of distribution mean that a deficit can occur at the same time in a certain geographical or social area, especially when biological objects like food, fuel, housing and clothes are concerned.

Supplies in the autotelic area are particularly enormous: there are generally far more games and dances, sports and athletics, 'refined arts', beautiful clothes, and all manner of undemanding 'entertainment' on tap in the modern urban environment than what the mind and psycho-physical receptor system can safely handle – not to mention financial packages to ease consumption.

Something similar affects the possibility of social relations – but here we run into classification problems. What in a given instance is object, and what is own ability? What is object lack and what is own lack? A popular technique involves projecting one's inadequacy onto the outside world; it's 'people' who don't cut the mustard. Occasionally, this apparent inability to cut the mustard happens to be true, and a useful piece of social defence withers on the vine for lack of an appropriate object – or possibly provokes aggression. Repatriated soldiers, for instance, are vulnerable to the dangers of such 'vacuum shocks'.

Where should the modern person go to find objects for their *metaphysical* needs? The inquiring mind, which seeks them in the world of experience, could come to suffer object deficiency. That said, even empirically inclined people could find a more or less stable object in a mix of experience and imagination (impressions, guesses, beliefs, etc.) such as, for instance, 'the triumph of good over evil', 'eternal peace', 'happiness and health for all', etc., as achievable future goals providing the effort is unselfish. In which case the 'need for meaning' will be satisfied. These objects (or as others would say: object surrogates) can be weighty and venerable – or they can descend into the purest parody (life missions as metaphysical concerns). A young bookbinder declared in 1953 that he would dedicate his life to raising the Norwegian bookbinding industry to and preferably above the level of the rest of the Nordic countries. If he succeeded, his life would have had 'as much meaning he had use for'.

The religiously minded believer would, if he belongs to an organized group (a congregation) find his object ready-made in the group's confession of faith. As a rule, objects consist of a state of grace after death.

3. Besides the needs and their objects, there are the *aids*, which today are multiple and variegated. But when the money economy supplanted the natural economy, *money* became the universal medium of exchange, so much so that many see money today as having intrinsic value. The mammoth hunter has metamorphosed into a white collar worker, he doesn't set off on a hunting sortie before lunch, for half of what he earns in a day he can visit the corner shop choose between fish from the sea, reindeer from the mountains, fowl from the forests, pork from the bacon curer and even whale meat. As temporary object, money also provides access to a wide array of mechanical aids, in farming, fishing, mining, manufacturing and daily life. At the same time, many of the main helpmeets of the past have gone out of circulation, more or less, such as personal characteristics like ingenuity and muscle power. The list re-emerges however in business, diplomacy and war. Most early biological needs have been transformed into social needs, society has largely assumed responsibility for the individual's protection, in return for him giving up corresponding areas of his liberty (judiciary, police, health service, etc., 'community care and support'). Money turns up everywhere, a fortune can give a person the social standing his personal attributes would otherwise not have earned him. In the autotelic area, money is the magic word, and in certain metaphysical systems it can even reinforce the sense of metaphysical confidence (alms, votive offerings, testamentary gifts to temple buildings, etc.). Technology is also put to use in the service of metaphysics (mission planes, evangelical readings on gramophone records, water-powered or electrically driven prayer wheels).

4. It is not always easy to tell between tools and *external conditions*, or, indeed, worth the while. One arranges the conditions as well as one can, and takes them into consideration whenever the tools are needed. Road conditions can be predicted, the daily trodden landscape is cultivated and standardized (bridges, roads, quays, airports), dangerous animals are exterminated. But bacteria hold the fort in many places, an ox or horse can run amok, earthquakes, hurricanes, volcanoes, blizzards, fires can still upset the neatest calculation. Social and economic conditions in a society will, as far as the individual 'little man' is concerned, be compartmentalized with 'external conditions', indeed, even certain psychological factors in himself. War, which is a human creation, comes without 'anyone wanting or

wishing it'. If the impact of civilization on the environment is beneficial (is an advantage with regard to the issue's enumerated, overall interest front) or disadvantage is the subject of regular debate among observers of the culture. One returns in one's holidays to the simple life in order to recharge one's batteries. The aeroplane has 'ruined' the ultimate adventure playgrounds for polar explorers and others searching for 'new land'.

5. The risk from acting to satisfy interests is largely mitigated. However, one can still risk damage to one's health and loss of social esteem. Psychological balance is increasingly vulnerable to critical events (the rising incidence of heart attacks and nervous disorders). Economic liability is a constant companion. In the autotelic sector, everyone is aware of the dangers of drinking, smoking and taking other stimulants to excess. Less obvious is the emergence of impersonal individual and 'nothing people' created by the constant barrage of 'entertainment' from cinemas, the 'coloured press', radio, etc. and their influence on latent proclivities.

6. At all times, a certain appeal issues forth to groups' and individuals' *opinions*, 'pattern of behaviour' as it is called nowadays. Attitudinal health has ensconced itself as a new, independent enterprise. An opinion can return and affect an individual's interest status through the opinion of outsiders. Hence, societies are largely saturated with a norm to show that 'optimism', whatever the circumstances, will never give in, never give up, never mope. And above all, never show despair or failures in public. On the other hand, there is an 'obligation' to 'show an interest' above all in current aspects of art and literature, politics and science, 'keep abreast' as they say. In many situations, loyalty to the groups to which one belongs, to the group's 'common interests', is required, despite the occasional failure in its observation.

Notes

1 The poem *Krig* (*War*) was published under the pseudonym of Per Fjeld, 15 July 1929.
2 This is a central point in the critique by system theoreticians of the mechanistic tradition in science (which, among other things, requires all processes to be reversible and that changes are not necessarily qualitative), of traditional Darwinism's accentuating of chance, and also of the analytical and quantifying social planning systems and economies which continue to dominate the field.
3 Which C.H. Waddington calls homeorhesos, to distinguish it from homeostasis.

6

THE DEEP ECOLOGICAL MOVEMENT

Some philosophical aspects

Arne Næss

(THIS TEXT WAS ORIGINALLY PUBLISHED IN 1993 IN S. ARMSTRONG AND R. BOTZLER (EDS): *ENVIRONMENTAL ETHICS: DIVERGENCE AND CONVERGENCE*. NEW YORK: MCGRAW-HILL)

Arne Næss (b. 1912) is head of the Philosophy Department at the University of Oslo, Norway. Næss first coined the term deep ecology in 1974, and has written extensively on the topic since that time.

Deep ecology is a movement designed to encourage people to question more deeply the fundamental resuppositions underlying the dominant economic approach of Western society in terms of value priorities, philosophy, and religion. Næss points out that deep ecology is not a proper academic philosophy, nor an institutionalized set of ideas as found in a religion or ideology. It is better characterized as a movement among people who share certain fundamental attitudes and beliefs, support a similar lifestyle, and agree on a variety of political issues. Næss presents eight points which help define a deep ecology perspective. He contrasts deep ecology with the shallow ecology position by their offering perspectives on pollution, use of natural resources, the human population explosion, cultural diversity and appropriate technology, land/sea use, and education and science. Næss identifies some of the underlying values or assumptions from which deep ecology can be derived in Christianity, Buddhism, oism, and Baha'i, as well as other philosophies. He then elaborates a well-defined personal philosophy 'Ecosophy T') which is founded on one ultimate norm: maximizing long-range, universal self-realization.

1. Deep ecology on the defensive

Increasing pressures for continued growth and development have placed the vast majority of environmental professionals on the defensive . . .

If professional ecologists persist in voicing their value priorities, their jobs are often in danger, or they tend to lose influence and status among those who are in charge of overall policies.[1] Privately, they admit the necessity for deep and far-ranging changes, but they no longer speak out in public. As a result, people deeply concerned about ecology and the environment feel abandoned and even betrayed by the 'experts' who work within the 'establishment'.

In ecological debates, many participants know a lot about particular conservation policies in particular places, and many others have strong views concerning fundamental philosophical questions of environmental ethics, but only a few have both qualities. When these people are silent, the loss is formidable . . .

2. A call to speak out

What I am arguing for is this: even those who completely subsume ecological policies under the narrow ends of human health and well-being cannot attain their modest aims, at least not fully, without being joined by the supporters of deep ecology. They need what these people have to contribute, and this will work in their favour more often than it will work against them. Those in charge of environmental policies, even if they are resource-oriented (and growth tolerating?) decision-makers, will increasingly welcome, if only for tactical and not fundamental reasons, what deep ecology supporters have to say. Even though the more radical ethic may seem nonsensical or untenable to them, they know that its advocates are, in practice, doing conservation work that sooner or later must be done. They concur with the practice even though they operate from diverging theories. The time is ripe for professional ecologists to break their silence and express their deepest concerns more freely. A bolder advocacy of deep ecological concerns by those working within the shallow, resource-oriented environmental sphere is the best strategy for regaining some of the strength of this movement among the general public, thereby contributing, however modestly, to a turning of the tide.

. . . Conservation strategies are more eagerly implemented by people who love what they are conserving, and who are convinced that what they love is intrinsically loveable. Such lovers will not want to hide their attitudes and values, rather they will increasingly give voice to them in public. They possess a genuine ethics of conservation, not merely a tactically useful instrument for human survival.

In short, environmental education campaigns can fortunately combine human-centred arguments with a practical environmental ethic based on either a deeper and more fundamental philosophic or religious perspective, and on a set of norms resting on intrinsic values. But the inherent strength of this overall position will be lost if those who work professionally on environmental problems do not freely give testimony to fundamental norms . . .

3. What is deep ecology?

The phrase 'deep ecology movement' has been used up to this point without trying to define it. In what follows, a set of principles or key terms and phrases (or a platform), agreed upon by George Sessions and myself, are tentatively proposed as basic to deep ecology.[2] More accurately, the sentences have a double function. They are meant to express important points which the great majority of supporters accept, implicitly or explicitly, at a high level of generality. Furthermore, they express a proposal to the effect that those who solidly reject one or more of these points should not be viewed as supporters of deep ecology. This might result because they are supporters of a shallow (or reform) environmental movement or rather they may simply dislike one or more of the eight points for semantical or other reasons. But they may well accept a different set of points which, to me, has roughly the same meaning, in which case I shall call them supporters of the deep ecology movement, but add that they *think* they disagree (maybe Henryk Skolimowski is an example of the latter). The eight points are:

1. The well-being and flourishing of human and non-human Life on Earth have value in themselves (synonyms: intrinsic value, inherent worth). These values are independent of the usefulness of the non-human world for human purposes.
2. Richness and diversity of life forms contribute to the realization of these values and are also values in themselves.
3. Humans have no right to reduce this richness and diversity except to satisfy vital needs.
4. The flourishing of human life and cultures is compatible with a substantially smaller human population. The flourishing of non-human life *requires* a smaller human population.
5. Present human interference with the non-human world is excessive, and the situation is rapidly worsening.
6. Policies must therefore be changed. These policies affect basic economic, technological and ideological structures. The resulting state of affairs will be deeply different from the present.
7. The ideological change will be mainly that of appreciating life quality (dwelling in situations of inherent value) rather than adhering to an increasingly higher standard of living. There will be a profound awareness of the difference between bigness and greatness.
8. Those who subscribe to the foregoing points have an obligation directly or indirectly to try to implement the necessary changes.

Comments on the basic principles

RE (1): This formulation refers to the biosphere, or more professionally, to the ecosphere as a whole (this is also referred to as 'ecocentrism'). This includes individuals, species, populations, habitat, as well as human and non-human cultures. Given our current knowledge of all-pervasive intimate relationships, this implies a fundamental concern and respect.

The term 'life' is used here in a more comprehensive non-technical way also to refer to what biologists classify as 'non-living': rivers (watersheds), landscapes, ecosystems. For supporters of deep ecology, slogans such as 'let the river live' illustrate this broader usage so common in many cultures.

Inherent value, as used in (1), is common in deep ecology literature, e.g., 'The presence of inherent value in a natural object is independent of any awareness, interest, or appreciation of it by any conscious being'.[3]

RE (2): The so-called simple, lower, or primitive species of plants and animals contribute essentially to the richness and diversity of life. They have value in themselves and are not merely steps toward the so-called higher or rational life forms. The second principle presupposes that life itself, as a process over evolutionary time, implies an increase of diversity and richness . . .

RE (3): The term 'vital need' is deliberately left vague to allow for considerable latitude in judgment. Differences in climate and related factors, together with differences in the structures of societies as they now exist, need to be taken into consideration.

RE (4): People in the materially richest countries cannot be expected to reduce their excessive interference with the non-human world overnight. The stabilization and reduction of the human population will take time. Hundreds of years! Interim strategies need to be developed. But in no way does this excuse the present complacency. The extreme seriousness of our current situation must first be realized. And the longer we wait to make the necessary changes, the more drastic will be the measures needed. Until deep changes are made, substantial decreases in richness and diversity are liable to occur: the rate of extinction of species will be 10 or 100 or more times greater than in any other short period of earth history.

RE (5): This formulation is mild. For a realistic assessment, see the annual reports of the World Watch Institute in Washington, DC.

The slogan of 'non-interference' does not imply that humans should not modify some ecosystems, as do other species. Humans have modified the earth over their entire history and will probably continue to do so. At issue is the *nature and extent* of such interference. The per capita destruction of wild (ancient) forests and other wild ecosystems has been excessive in rich countries; it is essential that the poor do not imitate the rich in this regard.

The fight to preserve and extend areas of wilderness and near-wilderness ('free Nature') should continue. The rationale for such preservation should focus mainly on the ecological functions of these areas (one such function: large wilderness areas are required in the biosphere for the continued evolutionary speciation of plants and animals). Most of the present designated wilderness areas and game reserves are not large enough to allow for such speciation.

RE (6): Economic growth as it is conceived of and implemented today by the industrial states is incompatible with points (1)–(5). There is only a faint resemblance between ideal sustainable forms of economic growth and the present policies of industrial societies.

Present ideology tends to value things because they are scarce and because they have a commodity value. There is prestige in vast consumption and waste (to mention only several relevant factors).

Whereas 'self-determination', 'local community' and 'think globally, act locally', will remain key terms in the ecology of human societies, nevertheless the implementation of deep changes requires increasingly global action: action across borders.

Governments in Third World countries are mostly uninterested in deep ecological issues. When institutions in the industrial societies try to promote ecological measures through Third World governments, practically nothing is accomplished (e.g. with problems of desertification). Given this situation, support for global action through non-governmental international organizations becomes increasingly important. Many of these organizations are able to act globally 'from grassroots to grassroots' thus avoiding negative governmental interference.

Cultural diversity today requires advanced technology, that is, techniques that advance the basic goals of each culture. So-called soft, intermediate and alternative technologies are steps in this direction.

RE (7): Some economists criticize the term 'quality of life' because it is supposedly vague. But, on closer inspection, what they consider to be vague is actually the non-quantifiable nature of the term. One cannot quantify adequately what is important for the quality of life as discussed here, and there is no need to do so.

RE (8): There is ample room for different opinions about priorities: what should be done first; what next? What is the most urgent? What is clearly necessary to be done, as opposed to what is highly desirable but not absolutely pressing? The frontier of the environmental crisis is long and varied, and there is a place for everyone.

The above formulations of the eight points may be useful to many supporters of the deep ecology movement. But some will certainly feel that they are imperfect, even misleading. If they need to formulate in a few words what is basic to deep ecology, then they will propose an alternative set of sentences. I shall of course be glad to refer to them as alternatives. There ought to be a measure of diversity in what is considered basic and common.

Why should we call the movement 'the deep ecological movement'?[4] There are at least six other designations which cover most of the same issues: 'Ecological Resistance', used by John Rodman in important discussions; 'The New Natural Philosophy' coined by Joseph Meeker; 'Eco-philosophy', used by Sigmund Kvaloy and others to emphasize (1) a highly critical assessment of the industrial growth societies from a general ecological point of view, and (2) the ecology of the human species; 'Green Philosophy and Politics' (while the term 'green' is often used in Europe, in the United States 'green' has a misleading association with the rather 'blue' Green agricultural revolution); 'Sustainable Earth Ethics', as used by G. Tyler Miller; and 'Ecosophy' (eco-wisdom), which is my own favourite term. Others could be mentioned as well.

And so, why use the adjective 'deep'? This question will be easier to answer after the contrast is made between shallow and deep ecological concerns. 'Deep ecology' is not a philosophy in any proper academic sense, nor is it institutionalized as a religion or an ideology. Rather, what happens is that various persons come together in campaigns and direct actions. They form a circle of friends supporting the same kind of lifestyle which others may think to be 'simple', but which they themselves see as rich and many-sided. They agree on a vast array of political issues, although they may otherwise support different political parties. As in all social movements, slogans and rhetoric are indispensable for ingroup coherence. They react together against the same threats in a predominately non-violent way. Perhaps the most influential participants are artists and writers who do not articulate their insights in terms of professional philosophy, expressing themselves rather in art or poetry. For these reasons, I use the term 'movement' rather than 'philosophy'. But it is essential that fundamental attitudes and beliefs are involved as part of the motivation for action.

4. Deep versus shallow ecology

A number of key terms and slogans from the environmental debate will clarify the contrast between the shallow and the deep ecology movements.[5]

Pollution

Shallow approach: Technology seeks to purify the air and water and to spread pollution more evenly. Laws limit permissible pollution. Polluting industries are preferably exported to developing countries.

Deep approach: Pollution is evaluated from a biospheric point of view, not focusing exclusively on its effects on human health, but rather on life as a whole, including the life conditions of every species and system . . .

The priority is to fight the deep causes of pollution, not merely the superficial, short-range effects. The Third and Fourth World countries cannot afford to pay the total costs of the war against pollution in their regions; consequently they require the assistance of the First and Second World countries. Exporting pollution is not only a crime against humanity, it is a crime against life in general.

Resources

Shallow approach: The emphasis is upon resources for humans, especially for the present generation in affluent societies. On this view, the resources of the earth belong to those who have the technology to exploit them. There is confidence that resources will not be depleted because, as they get rarer, a high market price will conserve them, and substitutes will be found through technological progress. Further, plants, animals and natural objects are valuable only as resources for humans. If no human use is known, or seems likely ever to be found, it does not matter if they are destroyed.

Deep approach: The concern here is with resources and habitats for all life forms for their own sake. No natural object is conceived of solely as a resource. This leads, then, to a critical evaluation of human modes of production and consumption . . . From a deep perspective, there is an emphasis upon an ecosystem approach rather than the consideration merely of isolated life forms or local situations. There is a long-range maximal perspective of time and place.

Population

Shallow approach: The threat of (human) 'overpopulation' is seen mainly as a problem for developing countries. One condones or even applauds population increases in one's own country for short-sighted economic, military, or other reasons; an increase in the number of humans is considered as valuable in itself or as economically profitable. The issue of an 'optimum population' for humans is discussed without reference to the question of an 'optimum population' for other life forms . . . A long-term substantial reduction of the global human population is not seen to be a desirable goal. In addition, the right is claimed to defend one's borders against 'illegal aliens', regardless of what the population pressures are elsewhere.

Deep approach: It is recognized that excessive pressures on planetary life stem from the human population explosion. The pressure stemming from the industrial societies is a major factor, and population reduction must have the highest priority in those societies.

Cultural diversity and appropriate technology

Shallow approach: Industrialization of the Western industrial type is held to be the goal of developing countries. The universal adoption of Western technology is held to be compatible with cultural diversity, together with the conservation of the positive elements (from a Western perspective) of present non-industrial societies. There is a low estimate of deep cultural differences in non-industrial societies which deviate significantly from contemporary Western standards.

Deep approach: Protection of non-industrial cultures from invasion by industrial societies. The goals of the former should not be seen as promoting lifestyles similar to those in the rich countries. Deep cultural diversity is an analogue on the human level to the biological richness and diversity of life forms. A high priority should be given to cultural anthropology in general education programmes in industrial societies.

There should be limits on the impact of Western technology upon present existing non-industrial countries and the Fourth World should be defended against foreign domination. Political and economic policies should favour subcultures within industrial societies. Local, soft technologies should allow for a basic cultural assessment of any technical innovations, together with freely expressed criticism of so-called advanced technology when this has the potential to be culturally destructive.

Land and sea ethics

Shallow approach: Landscapes, ecosystems, rivers, and other whole entities of nature are conceptually cut into fragments, thus disregarding larger units and comprehensive gestalts. These fragments are regarded as the properties and resources of individuals, organizations or states. Conservation is argued in terms of 'multiple use' and 'cost/benefit analysis'. The social costs and long-term global ecological costs of resource extraction and use are usually not considered. Wildlife management is conceived of as conserving nature for 'future generations of humans'. Soil erosion or the deterioration of ground water quality, for example, is noted as a human loss, but a strong belief in future technological progress makes deep changes seem unnecessary.

Deep approach: The earth does not belong to humans. For example, the Norwegian landscapes, rivers, flora and fauna, and the neighbouring sea are not the property of Norwegians. Similarly, the oil under the North Sea or anywhere else does not belong to any state or to humanity. And the 'free nature' surrounding a local community does not belong to the local community.

Humans only inhabit the lands, using resources to satisfy vital needs. And if their non-vital needs come in conflict with the vital needs of non-humans, then humans should defer to the latter. The ecological destruction now going on will not be cured by a technological fix. Current arrogant notions in industrial (and other) societies must be resisted.

Education and the scientific enterprise

Shallow approach: The degradation of the environment and resource depletion requires the training of more and more 'experts' who can provide advice concerning how to continue combining economic growth with maintaining a healthy environment. We are likely to need an increasingly more dominating and manipulative technology to 'manage the planet' when global economic growth makes further environmental degradation inevitable. The scientific enterprise must continue giving priority to the 'hard sciences' (physics and chemistry). High educational standards with intense competition in the relevant 'tough' areas of learning will be required.

Deep approach: If sane ecological policies are adopted, then education should concentrate on an increased sensitivity to non-consumptive goods, and on such consumables where there is enough for all. Education should therefore counteract the excessive emphasis upon things with a price tag. There should be a shift in concentration from the 'hard' to the 'soft' sciences which stress the importance of the local and global cultures. The educational objective of the World Conservation Strategy ('building support for conservation') should be given a high priority, but within the deeper framework of respect for the biosphere.

In the future, there will be no shallow environmental movement if deep policies are increasingly adopted by governments, and thus no need for a special deep ecological social movement.

5. But why a 'deep' ecology?

The decisive difference between a shallow and a deep ecology, in practice, concerns the willingness to question, and an appreciation of the importance of questioning, every economic and political policy in public. This questioning is both 'deep' and public. It asks 'why' insistently and consistently, taking nothing for granted!

Deep ecology can readily admit to the practical effectiveness of homocentric arguments:

> It is essential for conservation to be seen as central to human interests and aspirations. At the same time, people – from heads of state to the members of rural communities – will most readily be brought to demand conservation if they themselves recognize the contribution of conservation to the achievement of their needs as perceived by them, and the solution of their problems, as perceived by them.[6]

There are several dangers in arguing solely from the point of view of narrow human interests. Some policies based upon successful homocentric arguments turn out to violate or unduly compromise the objectives of deeper argumentation. Further, homocentric arguments tend to weaken the motivation to fight for necessary social change, together with the willingness to serve a great cause. In addition, the complicated arguments in human-centered conservation documents such as the World Conservation Strategy go beyond the time and ability of many people to assimilate and understand. They also tend to provoke interminable technical disagreements among experts. Special interest groups with narrow short-term exploitive objectives, which run counter to saner ecological policies, often exploit these disagreements and thereby stall the debate and steps toward effective action.

Writers with the deep ecology movement try to articulate the fundamental presuppositions underlying the dominant economic approach in terms of value priorities, philosophy and religion. In the shallow movement, questioning and argumentation come to a halt long before this. The deep ecology movement is therefore 'the ecology movement which questions deeper'. A realization of the deep changes which are required, as outlined in the deep ecology eight-point platform (discussed in section 3 above) makes us realize the necessity of 'questioning everything' . . .

6. Deep ecology illustrated as a derivational system

Underlying the eight tenets or principles presented in section 3, there are even more basic positions and norms which reside in philosophical systems and in various world religions. Schematically (Figure 6.1) we may represent the total views logically implied in the deep ecology movement by streams of derivations from the most fundamental norms and descriptive assumptions (level 1) to the particular decisions in actual life situations (level 4) . . .

As we dig deeper into the premises of our thinking, we eventually stop. Those premises we stop at are our ultimates. When we philosophize, we all stop at different places. But we all use premises which, for us, are ultimate. They belong to level 1 in the diagram. Some will use a sentence like 'Every life form has intrinsic value' as an ultimate premise, and therefore place it at level 1. Others try, as I do, to conceive of it as a conclusion based on a set of premises. For these people, this sentence does not belong to level 1. There will be different ecosophies corresponding to such differences.

Obviously, point 6 of the eight-point deep ecology tenets (see section 3) cannot belong to level 1 of the diagram. The statement 'there must be new policies

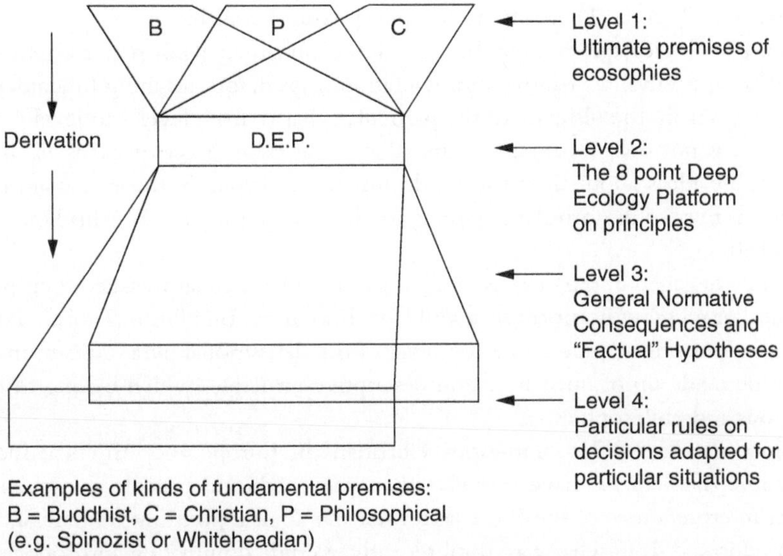

FIGURE 6.1 The apron diagram

affecting basic economic structures' needs to be justified. If no logical justification is forthcoming, why not just assert instead that ecologically destructive 'business as usual' economic policies should continue? In the diagram I have had ecosophies as ultimate premises in mind at level 1. None of the eight points of the deep ecology principles belong at the ultimate level; they are derived as conclusions from premises at level 1.

Different supporters of the deep ecology movement may have different ultimates (level l), but will nevertheless agree about level 2 (the eight points). Level 4 will comprise concrete decisions in concrete situations which appear as conclusions from deliberations involving premises at levels 1 to 3. An important point: supporters of the deep ecology movement act from deep premises. They are motivated, in part, from a philosophical or religious position.

7. Multiple roots of the deep ecology principles

The deep ecology movement seriously questions the presuppositions of shallow argumentation. Even what counts as a rational decision is challenged, because what is 'rational' is always defined in relation to specific aims and goals. If a decision is rational in relation to the lower level aims and goals of our pyramid, but not in relation to the highest level, then this decision should not be judged to be rational. This is an important point! If an environmentally oriented policy decision is not linked to intrinsic values or ultimates, then its rationality has yet to be determined. The deep ecology movement connects rationality with a set of philosophical or religious foundations. But one cannot expect the ultimate premises to constitute rational conclusions. There are no 'deeper' premises available.

Deep ecological questioning thus reveals the fundamental normative orientations of differing positions. Shallow argumentation stops before reaching fundamentals, or it jumps from the ultimate to the particular; that is, from level 1 to level 4.

But it is not only normative claims that are at issue. Most (perhaps all) norms presuppose ideas about how the world functions. Typically the vast majority of assertions needed in normative systems are descriptive (or factual). This holds at all the levels . . .

Those engaged in the deep ecology movement have so far revealed their philosophical or religious homes to be mainly in Christianity, Buddhism, Taoism, Baha'i, or in various philosophies. The top level of the derivational pyramid can, in such cases, be made up of normative and descriptive principles which belong to these religions and philosophies.

Since the late 1970s, numerous Christians in Europe and America, including some theologians, have actively taken part in the deep ecology movement. Their interpretations of the Bible, and their theological positions in general, have been reformed from what was, until recently, a crude dominating anthropocentric emphasis.

There is an intimate relationship between some forms of Buddhism and the deep ecology movement. The history of Buddhist thought and practice, especially the principles of non-violence, non-injury, and reverence for life, sometimes makes it easier for Buddhists to understand and appreciate deep ecology than it is for Christians, despite a (sometimes overlooked) blessedness which Jesus recommended in peace-making. I mention Taoism chiefly because there is some basis for calling John Muir a Taoist, for instance, and Baha'i because of Lawrence Arturo.

Ecosophies are not religions in the classical sense. They are better characterized as *general* philosophies, in the sense of total views, inspired in part by the science of ecology. At level 1, a traditional religion may enter the derivational pyramid through a set of normative and descriptive assumptions which would be characteristic of contemporary interpretations (hermeneutical efforts) of that religion.

Supporters of the deep ecology movement act in contemporary conflicts on the basis of their fundamental beliefs and attitudes. This gives them a particular strength and a joyful expectation or hope for a greener future. But, naturally, few of them are actively engaged in a systematic verbal articulation of where they stand.

8. Ecosophy T as an example of a deep ecological derivational system

I call the ecosophy I feel at home with 'Ecosophy T' (Figure 6.2). Ecosophy T has only one ultimate norm: 'Self-realization!' I do not use this expression in any narrow, individualistic sense. I want to give it an expanded meaning based on the distinction between a large comprehensive Self and a narrow egoistic self as conceived of in certain Eastern traditions of *atman*.[7] This large comprehensive

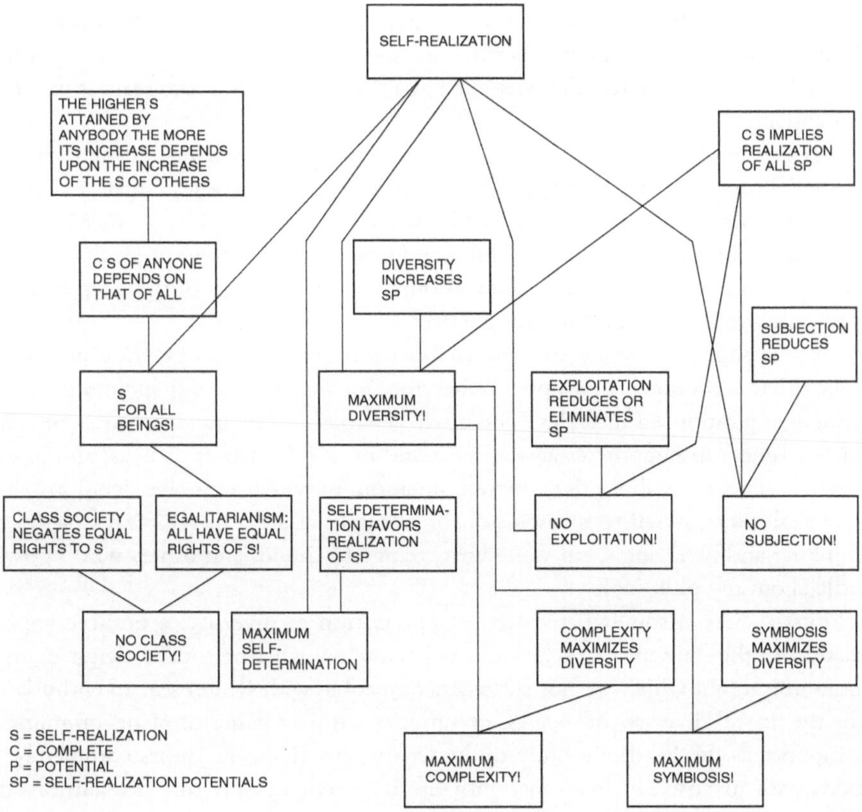

FIGURE 6.2 Ecosophy T diagram

Self (with a capital 'S') embraces all the life forms on the planet (and elsewhere?) together with their individual selves *(jivas)*. If I were to express this ultimate norm in five words, I would say: 'Maximize (long-range, universal) Self-realization!' Another more colloquial way to express this ultimate norm would be to say 'Live and let live!' (referring to all of the life forms and natural processes on the planet). If I had to give up the tem fearing its inevitable misunderstanding, I would use the term 'universal symbiosis'. 'Maximize Self-realization!' could, of course, be misinterpreted in the direction of colossal ego trips. But 'Maximize symbiosis!' could be misinterpreted in the opposite direction of eliminating individuality in favour of collectivity.

Viewed systematically, not individually, maximum Self-realization implies maximizing the manifestations of all life. So next I derive the second term, 'Maximize (long range, universal) diversity!' A corollary is that the higher the levels of Self-realization attained by any person, the more any further increase depends upon the Self-realization of others. Increased self-identity involves increased identification with others. 'Altruism' is a natural consequence of this identification.

This leads to an hypothesis concerning an inescapable increase of identification with other beings when one's own Self-realization increases. As a result, we increasingly see ourselves in other beings, and others see themselves in us. In this way, the Self is extended and deepened as a natural process of the realization of its potentialities in others.

By universalizing the above, we can derive the norm, 'Self-realization for every being!' From the norm, 'Maximize diversity!' and a hypothesis that maximum diversity implies a maximum of symbiosis, we can derive the norm 'Maximize symbiosis!' Further, we work for life conditions such that there is a minimum of coercion in the lives of others. And so on![8] The eight points of the deep ecology platform are derived in a fairly simple way.

A philosophy as a world view inevitably has implications for practical situations. Like other ecosophies, Ecosophy T therefore moves on, without apologies, to the concrete questions of lifestyles. These will obviously show great variation because of differences in hypotheses about the world in which each of us lives, and in the 'factual' statements about the concrete situations in which we make decisions.

I shall limit myself to a discussion of a couple of areas in which my 'style' of thinking and behaving seem somewhat strange to friends and others who know a little about my philosophy.

First, I have a somewhat extreme appreciation of diversity; a positive appreciation of the existence of styles and behaviour which I personally detest or find nonsensical (but which are not clearly incompatible with symbiosis): an enthusiasm for the 'mere' diversity of species, or varieties within a genus of plants or animals: I support, as the head of a philosophy department, doctrinal theses completely at odds with my own inclinations, with the requirement only that the authors are able to understand fairly adequately some basic features of the kind of philosophy I myself feel at home with; an appreciation of combinations of *seemingly* incompatible interests and behaviors, which makes for an increase of subcultures within industrial states and which might to some extent help future cultural diversity. So much for 'diversity!'.

Second, I have a somewhat extreme appreciation of what Kant calls 'beautiful actions' (good actions based on inclination), in contrast with actions which are performed out of a sense of duty or obligation. The choice of the formulation 'Self-realization!' is in part motivated by the belief that maturity in humans can be measured along a scale from selfishness to an increased realization of Self, that is, by broadening and deepening the Self, rather than being measured by degrees of dutiful altruism. I see joyful sharing and caring as a natural process of growth in humans.

Third, I believe that multifaceted high-level Self-realization is more easily reached through a lifestyle which is 'simple in means but rich in ends' rather than through the material standard of living of the average citizens of industrial states.

The simple formulations of the deep ecology platform and Ecosophy T are not meant primarily to be used among philosophers, but also in dialogues with the 'experts'. When I wrote to the 'experts' and environmental professionals personally, asking whether they accept the eight points of the platform, many answered positively in relation to most or all of the points. And this includes top people in the ministries of oil and energy! Nearly all were willing to let their written answers be widely published. It is an open question, however, as to what extent they will try to influence their colleagues who use only shallow argumentation. But the main conclusion to be drawn is moderately encouraging: there are views of the human/nature relationship, widely accepted among established experts responsible for environmental decisions, which require pervasive, substantial changes of present policies in favour of our 'living' planet, and these views are held not only on the basis of shortsighted human interests.

Notes

1. These problems are discussed further in Næss' keynote address to the second international Conference on Conservation Biology held at The University of Michigan in May 1985; published as 'Intrinsic Value: Will the Defenders of Nature Please Rise?', *Conservation Biology* (1986): 504–515.
2. The Deep Ecology principles (or platform) were agreed upon during a camping trip in Death Valley, California (April, 1984) and first published in George Sessions (ed.), *Ecophilosophy VI* newsletter (May, 1984). They have subsequently appeared in a number of publications.
3. Tom Regan, 'The Nature and Possibility of an Environmental Ethic', *Environmental Ethics* 3 (1981): 19–34, citation on p. 30.
4. I proposed the name, 'Deep, Long-Range Ecology Movement' in a lecture at the Third World Future Research conference in Bucharest in September, 1972. A summary of that lecture ('The Shallow and the Deep, Long-Range Ecology Movement') was published in *Inquiry* 16 (1973): 95–100. Within the deep ecology movement it is fairly common to use the term 'deep ecologist', whereas 'shallow ecologist', I am glad to say, is rather uncommon. Both terms may be considered arrogant and slightly misleading. I prefer to use the awkward, but more egalitarian expression 'supporter of the deep (or shallow) ecology movement', avoiding personification. Also, it is common to call deep ecology consistently anti-anthropocentric. This has led to misconceptions: see my 'A Defense of the Deep Ecology Movement', *Environmental Ethics* 5 (1983).
5. The 'shallow/deep' dichotomy is rough. Richard Sylvan has proposed a much more subtle classification; see his 'A Critique of Deep Ecology', *Discussion Papers in Environmental Philosophy,* RSSS, Australian National University, No. 12 (1985).
6. *World Conservation Strategy*, section 13 (concluding paragraph).
7. The term *atman* is not taken in its absolutistic senses (not as a permanent indestructible 'soul'). This makes it consistent with those Buddhist denials (the *avatman doctrine*) that the *atman* is to be taken in absolutist senses. Within the Christian tradition some theologians distinguish 'ego' and 'true self' in ways similar to these distinctions in Eastern religions. See the ecophilosophical interpretation of the gospel of Luke in Stephen Verney's *Onto the New Age* (Glasgow: Collins, 1976), pp. 33–41.
8. Many authors take some steps toward derivational structures, offering mild systematizations. The chapter 'Environmental Ethics and Hope' (in G. Tyler Miller, *Living in the Environment,* 3rd edn [Belmont: Wadsworth, 1983]) is a valuable start, but the derivational relations are unclear. The logic and semantics of simple models of normative systems are briefly discussed in my 'Notes on the Methodology of Normative Systems', *Methodology and Science* 10 (1977): 64–79. For a defence of the thesis that as soon as people assert anything at all, they assume a total view, implicitly involving an ontology, methodology, epistemology, and ethics, see my 'Reflections about Total Views', *Philosophy and Phenomenological Research* 25 (1964–5): 16–29. The best and wittiest

warning against taking systematizations too seriously is to be found in Soren Kierkegaard, *Concluding Unscientific Postscript*.

For criticism and defence of my fundamental norm ('Self-realization'), together with my answer, see *In Sceptical Wonder: Essays in Honor of Arne Næss* (Oslo: University Press, 1982). My main exposition of Ecosophy T was originally offered in the Norwegian work, *Okologi, samfunn og livsstil* (Oslo: University Press, 5th edn, 1976). Even there, the exposition is sketchy. [Editor's note: Næss' Norwegian book has been revised and reissued as Arne Næss, *Ecology, Community and Lifestyle*, trans. and ed. David Rothenberg (Cambridge: Cambridge University Press, 1989)].

7

GAIA VERSUS SERVOGLOBE

Sigmund Kvaløy Setereng

Introduction

The core content of this chapter was originally given as a talk at an international conference on the 'Gaia theory', arranged by ECOROPA in Cornwall, England in 1987.[1] My interest in globalized microprocessor control goes back, however, to the early 1950s, initially embracing the concept with science fiction-inspired enthusiasm. Over the years that attitude changed radically, and it finally came to a head during my stay at Columbia University, New York, in 1967. I was there to study and produce something on 'The aesthetic and communicational problems of electronic and computer-generated music'. New York is also the Mecca of jazz music, and I was deeply struck by the contrast between the two forms: constructivism versus improvisation; they appeared finally as two different worlds, and that became the starting point for the development of the concepts 'complication' and 'complexity' (defined below).[2]

I left New York and the Ecophilosophical Group was formed at the University of Oslo in 1969: we watched the expansion of the European Economic Community (EEC) – inspiring the concept 'Industrial Growth Society' (later designated ACID; see below), and in 1972, seven of us produced the book *Økopolitikk eller EEC?* (Ecopolitics or the EEC?), which sold 10,000 copies – to our surprise, since it was somewhat on the philosophical side. One of the books the group had discussed was Christopher Hodder-Williams' *A Fistful of Digits* (see bibliography), giving us the idea of 'Servoglobe'; he uses the designation 'Servex' – which we also used in the beginning, until we were threatened to be sued in court by the washing machine company SERVEX.

That made us invent the word Servoglobus (*Servoglobe*), which, anyway, turned out to cover the intended meaning better than the earlier term. Other analyses that helped form our own were those made by Jacques Ellul (*The Technological Society*),

Lewis Mumford (*The Myth of the Machine*), Peter Wessel Zapffe ('The Norwegian Apparatus Landscape'), Norway's earliest ecophilosopher, philosopher and sociologist Dag Østerberg (*Forståelsesformer* – 'Forms of Understanding'), as well as the more ecophilosophical parts of the British monthly *New Scientist* (for additional sources, see the bibliography).

Inspired by these, and my collaborators in the ecophilosophical group (Arne Vinje, Paul Hofseth, Ivar Mysterud, Per Gaarder, Nils Faarlund, Erling Amble, Jon Godal, Karl Georg Høyer – and Arne Næss and Dag Østerberg as important satellites to the group), still being politically naive but thinking we had a message of overwhelmingly serious significance, I approached members of the parliament around the mid-1970s as batches of new communication and surveillance systems were put into space (notably satellites etc. of the National Security Agency (NSA) – arguably the world's most secret global espionage organization). I was met with friendly nods but it quickly became clear that I would need a week or two of talking space to establish communication on the issue, and of course the parliamentarians' space available for me was ten minutes at the most.

Anyway, our book on Ecopolitics and the EEC has turned out to have a great predictive value, e.g. that the socio-economic process let loose by the EEC's founders was a self-strengthening spiral (a positive feedback affair) towards a union (the leading opinion-makers scoffed at this possibility in 1972, and, strangely, still do it today (1999), when the Union is more or less a fact).

In later years I have had the opportunity to meet and discuss the main ideas of this chapter with some of the authors who helped inspire it in the first place, like Hubert Dreyfus and Theodore Roszak in California (see bibliography), the Dalai Lama as well as Buddhist philosophers of Bhutan, and most prominently, Helena Norberg-Hodge, founder and director of the International Society for Ecology and Culture, as well as Edward Goldsmith, editor of *The Ecologist* – a leading ecophilosophical journal.

Of course I am mentioning the interaction with these people to try to convince the reader that the *Servoglobe* prediction is not just the product of a mind gone astray in the face of the contemporary turmoil. Anyway, Norway appears to me to be the country where society's computerization has been accepted with the greatest naivety. Now and then up through the passing years, returning again and again with inspiration from abroad, I have attempted to start a debate in my own country, using available opportunities in the various media; it always came to nothing. Our academic environment is small and we largely lack politically engaged generalists to face the specialists.

Lately, however, the law professors Thomas Mathiesen and Ståle Eskeland have tried to raise awareness of and an opinion against the Big Brother scheme of the European Union (Schengen, Europol, etc.). No political reaction seems to be forthcoming. Georg Apenes, the Director of Datatilsynet ('Data Watch') – a state institution that watches and can stop computer violation of privacy – is almost the only one in Norwegian society who manages to stir up a little debate now and then. None of these, however, appear to have gone into the psychosocial and ecopolitical impact of the computerization process in its global reach.

The present attempts at interdisciplinary team research at all four of the Norwegian universities are, if anything, positive towards computerization, revealing how this new trend lacks in depth. To argue that failure of insight is the substance of this chapter.

Summary of main points

First, 'Servoglobe' is a name given to a supra-nationally run 'global supercomputer' – an 'artificial intelligence' system coupled to a global network of information gathering and electronic data-processing systems, surveillance satellites, regional expert terminals, etc., semi-automatic in operation – created partly by economic globalization forces, and partly to serve mankind's survival in the face of a destructively simplified natural biosphere (Gaia) and the threat of political and social chaos.

Second, there are strong scientific, technological, economic, administrative and political forces at work on all sides of the various global conflict fronts to reach efficiency in repairing and managing the increasingly unmanageable mess created by Western Industrial Society – or rather: the Advanced Competitive-Industrial Dominion – ACID (to be defined). There have been various research projects in operation for more than thirty years to co-ordinate this effort through micro-processing networks of increasing sophistication. Various feedback loops are involved in this process, gradually producing a self-propelling impetus of such a force that it is hard to see what would stop it, even if there were any public awareness of what is happening. There is not. Projecting this development into the future, the end point is *Servoglobe*.

Third, part of the strength of this self-propelling process has its root in the impossibility of managing a natural dynamic 'system' as if it was a system of static machinery (the 'complex' as if it was 'complicated', see below), but: *these are the only means that science and technology have given us.* The result is a build-up of chaos-potential, and at an accelerating rate larger quantities and more refined versions of the mechanistic 'medicine' are applied. Instead of changing the medicine, which is beyond the horizon of ACID, a steadily greater effort is put into widening its application: the thinking is that 'there are loopholes in the systems' which must be closed to achieve a completely rational functioning of it. Among other things, human political activity is treated as 'sand in the machinery' that must be removed to obtain a smooth running. So are old borders between nations. *Servoglobe* is the 'Endlösung' to these problems.

Fourth, finally, nature – the biosphere, Gaia – must also be conceived by this thinking as too messy for rational management. After all, she is seen primarily as a collection of resources for human material needs. So, the impetus is strong towards a replacement of Gaia – i.e. the biosphere's natural self-regulation – by an artificial system of global compass.

Fifth, *Servoglobe*, however, is an impossible goal – a breakdown of the mechanistic control network will happen before that end-point is reached. The reason for this is the later mentioned lack of 'interface' between '*the complicated*' and '*the complex*'. These concepts are defined (this is the central part of the presentation, which everything else hangs on): Gaia is complex, ACID tends inherently towards perfected complication.

Sixth, ACID may go far in this grandiose attempt and, the further it gets, the more devastating will be the resulting chaos (Priogine is counterargued). To create awareness of this development is therefore an extremely vital project.

Central concepts

Computers large and small abound and proliferate exponentially. Even my own country, Norway, reputed as an exceptionally down-to-earth nation, has been hit by an almost frantic propaganda for their use in all possible corners of society, schools being the prime target. We must bring our people quickly into the computer age, otherwise we'll lose out in the international competition. Not only economically, but culturally too.

A decade ago we saw one TV programme after another telling us that we should watch the children – '*they* are not afraid!' 'Just try for a while, and you'll be convinced!' – and we were promised new jobs, as well as a new creativity and even liberation – from *angst,* from chaos, from war. By now, we are convinced; very little of the old fear remains. The media hardly ever publish any critical comment. Directly, and indirectly, in many ways, we are led to believe that computers and human brains belong to the same class of phenomena, reducing the 'interface' problems to something almost negligible – around the next corner the computer will exist as a natural extension of the human mind. Both to blue/red industrialists and green 'New Age'-followers, interconnected computers will solve our problems on a global scale.

For those of us who think that we are here in for the most consequential and unfathomable upheaval of society that human history ever witnessed, among other things, due to the speed and universality with which the computers take over vital functions, it's high time we sit down and try to grasp as well as we can the depth of what is happening. Time is, in my opinion, far overdue for green associations, like the intra-European ECOROPA, and The International Forum on Globalization, to do this, since what we see here is, in effect, an attempt to replace the naturally flowing ecosocial system with an artificially controlled, mechanistic scheme.

I will mention three concepts that to me have been useful on distinguishing various elements that the pro-computer advocates mix up to everybody's confusion. I label these concepts *complexity*, *complication* and *pseudo-complexity*; in the latter case I also, depending on the context, use the expressions 'Amusement diversity' and 'Disneyland Effect'. This handful of concepts have been helpful to me in clarifying, among other matters, how computers and living entities differ to the extent that they belong to different worlds. As will be seen, the concepts contain hypotheses about nature, society and machines.

By 'complexity' (CX) I mean the dynamic, irreversible, non-centrally self-steered, goal-directed, conflict-fertilized manifoldness of nature and – as a particularly refined and intricate version of that – the human mind/body entity.

By 'complication' (CC) I mean the static, reversible, externally and uni-centrally steered, standarized structure-intricacy of the machine. The computer is a particularly refined and intricate version of that.

'Pseudo-complexity' (PCX) is, like CC, a human design-product, but unlike CC it mimics CX in order to set up various arrangements and activities to keep people occupied in a diverse manner, through mass media, hobbies, tourism, schools, etc. so that it replaces complex challenges; it occupies people on the shallow level that is exemplified by the amusement park, i.e. without offering training or development that equips them better for creative interaction with nature and society. Environmental PCX often functions as a 'safety valve' for the inner urge towards complex integration with the larger world that every human being is born with; it offers, however, only the sort of interaction that leaves the personality without further development or maturation after the event.

CX can be described only by reference to *qualities* (where, however, quantitative perception and assessment is included as one segment of the spectrum of qualitative mediations; between the individual and her/his environment) – kinds and sorts of differentiations and border-crossings, shades and hues without fixed boundaries, dialectical build-ups and break-throughs, etc.

CC can, in any situation, be completely described by reference to the five mechanical parameters: height, breadth, depth, mass, and locomotion – modelled through *spatial* diagrams, mathematics and formal logic, and quantified according to fixed numerical scales.

CX is thought of as something existing, as being real – as concrete, imperfect, practice-related goings on in the human body-mind its world, and as something already given and presented when 'consciousness wakes up' and starts to formulate concepts. It's in large measure pre-intellectual.

CC is a product of the human intellect and manifests itself as abstract model-collections, describing ideal, theoretical, perfect machine-like structures. Since Pythagoras and Plato, and modernized by Descartes, CC has largely been taken to express the real world,[3] and that is still dominating Western thought, making it easy to accept the computer as an extension of the living.

Time

To understand CX and CC and their difference, it is useful to apply two different time concepts, 'organic time' as a part-characteristic of CX and 'mechanical time' (or 'clock time') as the 'time' of CC. Organic time is the group of modes of change related to the various subjects and actors and groups of such in nature, it expresses itself through rhythm, expands and contracts, is created as part of natural processes, does not extend into the future, nothing moves through it – it is movement itself. Organic time is concrete; it has no existence separate from material-spiritual events ('being-processes' is one tentative word that might be useful to get away from the Cartesian split between matter and spirit). It cannot be 'objectively' measured, because any kind of measuring rod would change with it, but among humans it can be illustrated and talked about concretely, clearly, and with abundant communicability. Available to us for such communication are innumerable human sensual, emotional, aesthetic and ethic experiences ('I arrived earlier than my uncle, who was actually walking very fast – red-faced and puffing, but strong – despite his heart condition', etc.!).

Mechanical time is a human intellectual invention. It was historically prepared through the observation of the rhythmical pulses of day and night and the seasons, but came into consequential being as something different from organic time only with the invention of a reliable mechanical 'time-keeper' (a device that conserves time) – the clock. It is used to quantify natural time – or, rather, to substitute mechanical co-ordination for rhythmical living in approximation with the sun's rising and setting, tides in the sea, and various rhythmical body functions. Mechanical time is regarded as being the same for all natural processes; it's 'objective' not 'subjective', we speak as if we are 'moving through it', it is supposed to stretch out into the future (it is actually constituting the concept 'future' in Western civilization – a sheet of paper, a map, now a screen, on which the *planners* trace their lines), as if what has not yet happened is already 'there' 'some place' (it's embryonically also the basis for the belief in precognition and time travel – strengthened today through science fiction – where H.G. Wells' *The Time Machine* is the classic). It keeps us in captivity since it's a fenced-in path for our steps, but it is abstract – not dependent upon concrete events, and so lacks qualitative aspects. It is the basis for the notion of reversibility of processes, and it is the only time concept relevant to the construction of the physical computer and the programming of it as well.

Human culture as we know it, can hardly function without the co-ordination of mechanical time. The trouble is, however, that it has a tendency to hide natural time, and as *the dominant* time concept of industrialism it has made Western man almost lose awareness of natural time – a fact of deep consequence, including the notion that the human brain/mind relationship is a kind of computer/computer program ('software') system. Logically, mechanical time is a *spatial* parameter – it is the measuring rod for locomotion – shift of location – in geometrical co-ordinate systems. As said it psychologically tends to eliminate (natural, real) time. Only space is left. In accordance with this perspective, the Advanced Competitive-Industrial Digit-Society (ACID), the social system that – like acid – simplifies complex entities, is an extreme experiment in testing how far it is possible to go in spatializing the human universe, an attempt that started in classical Greece (Pythagoras, Parmenides, Plato, Xenon, etc.).

Two cultures – two time concepts

As an illustration of this historical development, the modern Western house – an engineering perfection of the Greek temple – has a smooth, geometrically perfect appearance, seemingly time-defiant. If a crack appears in its shining surface – an attack of organic time – it looks terrible because decay is supposed to be irrelevant to its conception. It is an expression of a 'stop-time aesthetic'. In contrast, if we take a look at – say – a Sherpa house in Nepalese Himalaya, it always appears 'unfinished', a creation that never reached 'its destined geometrical perfection'. This, however, is the modern Western perception of it – that of the visiting tourist or development aid expert. We stop at the entrance of the village and say: 'The poor ignorant Sherpas, they can't to a better job.'

But, from the traditional Sherpa point of view, the beauty and, intimately connected with that, the utility of his house may only be discovered if you settle down for a couple of generations, build such a house yourself, take responsibility for its daily care, live with the house instead of being its architect, repair it when (the frequent) need arises, add to it or subtract from it as the requirements of your family and your animals may be, etc. The modern Western building is given an adequate expression through a snapshot photo, while there's another modern Western device that might help you to an initial appreciation of the quality of a 'Sherpa type house', the cinematic film camera. You mount it on a tripod in front of the house and let it expose one frame every day for, say, 100 years. Finally you develop the film and let it run at normal cinematic speed. What will be revealed to you, is not a house in the Western sense, but an organic structure, its wall stones and roof material will be moving about and changing to it, the animal and human life around it will expand and contract, speed up and slow down, shift in kind and variety – all in subtle co-ordination with changes appearing in the house itself. You are viewing a total, complex process, where all elements in the environment and economic/social conditions prevailing in the locality, quickly and flexibly are mirrored in the house.

This is a house that is decaying every day, a fact which is *accepted* by the people that are part of this 'house-hold' – a material, pliable dough, responding to both the immediate and the long-term challenges of local life. The people of the house let it decay, always, to some extent, so that it can bud and sprout every next spring, but at any *moment* moving to be something new. This house, behaving like a living organism, is the expression of a culture that also accepts illness, age and death, that even finds security in accepting that nature is an insecure place, a process. Strengthening that, the Sherpas have a religion – Buddhism – founded on the premise that nothing is permanent. The result is that quality of self-reliance, resourcefulness, ability to improvise, and flexible cooperation which makes the Sherpas' help so attractive to Western mountaineers engaged in the world's most demanding ventures in wild nature, summit climbing in the Himalayas.[4]

Why wouldn't the computer be of any help here, but rather a hindrance? Because we are faced with a particularly clear case of Gaia's demands on natural, complex man. The challenges facing the Himalayan mountaineer are hardly ever complicated. But in principle, these demands are no different from, say, those facing someone bringing up a child – anywhere, even in the West (which tends, however, to 'complicate' that issue too). In both cases, complex growth is the challenge. Correspondingly, the computer may be useful in constructing and running a modern Western office building, while it is a total miss in a Sherpa village life process. And the office building, in order to stay smooth, as a daily sign of corporate reliability, continuously demands resources far beyond what the neighbourhood can give. The same goes for the Western urban system as a totality – and any attempt at keeping a complicated structure smooth!

Meaningful, complex work versus complicated employment in the Gaian perspective

There is, from what I have said, justification in using the expression 'real time' for natural time, since it is irreducibly concrete – movement itself as experienced by each individual, while mechanical time is an abstraction – a reducing of time to geometrical space. Under the heading 'a complicated system (like the computer is incapable of handling real time)', there is one aspect where 'natural man's' incompatibility with the computer is particularly consequential. To my thinking, the concept of '*meaningful work*' is central to understanding of the unfoldment of human personality, and I am here talking about something specifically human. It is the aspect of 'human-ness' that – before anything – is the product of man as an inventive socio-ecological creature. 'Meaningful work' – as I will define it – is 'human work', in contradistinction to 'mechanical work' or 'complicated employment'. Again: the complex versus the complicated.

I define *meaningful* – or *humanly appropriate work* (mfW – where 'mf' also stands for manifold-ness, i.e. complexity) as follows:

1. It is an activity *necessary* for the human being's material survival (through this, it possesses a direct, clear seriousness not shared by any other activity except defence against disruptive social conflict).
2. Its *fruit* of products (goods, services . . .) are such that do not damage but rather strengthens life (human society and nature) with no envisaged time limit – which means that Gaia is always – more or less consciously – on the horizon.
3. It poses such challenges that bring the potential *complexity* of capabilities – including the ability to improvise, i.e. alertness to organic time – in the human individual and her group to bloom.
4. It demands of its partakers the building of *solidarity and loyalty* and practical, flexible techniques for *cooperation*; and
5. In general it engages *children* (as the most important group – but any other, like old people, as well).

Years of active life filled with mfW establishes constant awareness of organic time, including personal, dialectical shifts – i.e. conflict-fertilized maturation.

Work roughly corresponding to this concept has as a matter of fact been the normal basis of society – its economy, education, artistic and political ability – worldwide and as far back as we are able to trace human history. *The first society to completely lack this basis* is ACID, and the computer perfects this state of affairs: it employs people, i.e. segments of them, to some extent, but robs them of work. A complex society, with complex process-attachments to Gaia entails complex persons! Sums of specialists or segment-persons won't do. To state very briefly one of the many lines of argument involved in the matter: a society built on the basis of mfW is a society that in principle thrives on meagre resources in energy and materials – while energy/materials abundance blocks its realization.

Contrary to ACIDic thinking (also in greener branches, New Age, 'Decentralised Regionalism!' etc.), micro-electronics presupposes an energy-intensive, international heavy-industry economy, factually accelerating the depletion of the world's resources: the word 'post-industrial' stands in my analysis for idealistic wishful thinking!

The main point here is that mankind's greater teacher is Gaia in her manifoldness, always up to something new; but she will fulfil that function only if her complex challenges are not replaced by complicated problem-tasks, posing as the easy way out – usually through the ready availability of specialists. (Wo)man is by birth pre-eminently gifted towards being a continuously creative Gaia-being in body-spirit, but that creativity is a complex affair, which as a *general, social* characteristic is brought to flower only in cultures that have their basic flow-structure determinated through meaningful work, posed as a challenge necessary for survival. This line of reasoning owes nothing to the 'Christian/Protestant work ethos' a generation or two back, but is founded on empirical observation, comparing how societies live and develop within various cultures. The point is that a human animal is free, even in the sense that it may 'choose' the easy way out, if its environment offers that, like ACID. Then it may expand in a pseudo-complex manner, or just 'go to pot', ashes-like.

I am here largely in keeping with E.F. Schumacher in the chapter 'Buddhist Economics' from the book *Small is Beautiful*, where he stresses the point and that to my mind has paramount importance, namely that 'man needs work *as much as* he needs food', and that 'man without work is in a desperate situation' – where the concept 'work' is closely related to mine (except for point 5 – the requirement of children's participation, which of course is indispensable if we want to restore to work the role as society's foundation).[5] Taking this point seriously constitutes an elementary step towards the much-sought-for Green Paradigm. In spite of its extremely well-entwined historical roots, it is a truly radical, concrete and forceful departure from ACID – one that we cannot possibly avoid if our course is to deserve the label 'social' or Gaia-directed.

In contradistinction to this, the computerized society model furthers isolated individualism – as it is spelled out, e.g. in Zbigniev Brzezinski's grand vision of a 'Technetronic Society',[6] *a* 'society' that in my analysis would be economically dependent on mechanistic schemes replacing organic processes producing an alienation to nature that also means human self-alienation. It offers an abstract world in replacement of a concrete, a complicated for a complex, machine-directed employment instead of work. It represents the last stage of a Western civilization that for centuries increasingly regarded work purely as a means towards an end – survival, and, if possible, affluence: work acquired meaning *only* from the end it was to serve – from itself only as a chance byproduct. The socialist workers' revolution happened within this historical context, and for that reason, partly, *meaningful* work – work in itself possessing the promise of human bodily/intellectual and emotional expansion – never became an important fighting issue to the socialist movement. If it had been the case, computerized employment would have had a much harder time in gaining labour's acceptance than we see today. Let's call this 'Karl Marx's blind spot'. Due to historical urgencies in his time, we can't blame him. His day required the red so much that it had to blot out the green. In my opinion, after thirty years of discussing work and social change on three continents (admittedly a short time – clockwise), Green's central key is meaningful work, as defined.

MfW requires (wo)man to accept insecurity as a normal state of affairs, and it trains the individual or the group to be broadly prepared for new challenges, to improvise – to be creatively one with organic time. That is why it shares a common departure with Buddhism, which in some of its historical branches represents the world's most radical eco-process philosophy. (This is *not* the case with the *Avatamsaka Sutra* story of *Indra's Net*, often quoted by New Age people. Instead, we have here to do with a sort of Far Eastern 'Spinozistic' eternalism, where everything reflects everything else in a perfect manner. The crystalline world net mentioned in the story is rather some sort of 'super-complication', inspired in its way by meditation, which at its deep level is quality-free[7].) In those branches of Buddhism where its originality is kept fresh *nothing* is permanent, everything is change and movement – the idea that anything is enduring, like even the most highly cherished objects, the human soul and God – is an illusion.

In keeping with this, but transposing Buddhism to our time and place, I define the human personality through *activity*, more or less creative, and the individual's activity as it ties in with other individuals' activity. The individual is here thought of as a process – its being is its activity, slow-moving and gradual or creative by leaps, only hazily distinguishable from other individuals' activity processes (unlike Indra's net, nothing is ever a copy, the world is full of mutually overlapping, reciprocally inspired individualities, new ones and new constellations forever emerging – human, animal and plant. This kind of reality explains why the scientific method, as we know it, can never grasp more than distorted fragments of the world). The 'other-individuals' may here be either animal or vegetable, or even non-animate – like a mountain, a desert or a river. Everything concretely nature-given is movement, perceivable through characteristic rhythms and modes of rhythm shifts, while the world of mechanistic science is static, mathematical and abstract (I do *not* say that mathematics is non-creative or useless!).

Even though new approaches may be appearing in modern microphysics, the way science is taught and practised is still mechanistic, which it must be anyway as long as it serves ACID. The computer is part of the latter, regardless of its vastness in electronic miniaturization. Either digital or analogue, it's still qualityless – complicated not complex: the computer functions on the basis of sharply defined boundaries, the individual it represents is either an abstract – binarily catalogued 'cat' *or* 'dog', never the dog-like cat that crossed my way as I left the farm to start on the complicated and complex journey to an ECOROPA meeting. A robot can never be any sort of open-ended, inventive activity-presence, at every stage redefining its character, as any member-being of concrete Gaia. Computer-locomotion is reversible – as long as it fulfils its expectations, Gaia's and man's activity is not.[8]

Ecology was at one time defined by the American ecophilosopher Paul Shephard as 'the subversive science'. Today students, by the hundred thousands, learn about 'nature' in front of a computer, displaying various 'simulations'. They learn about something complicated instead of something complex, and are hardly ever led on to the idea that there might be a difference. It's now an overdue and extremely important task for the ecopolitical movement and for ecophilosophical pedagogies to stir up a debate within the field of education in general about the practice of representing complexity as complication. A critical attitude to the computer would – as things have developed – have to be the central part of such a project. The computer draws the whole ACIDic development together – makes it finally crystal clear where it's all been leading us, what kind of a hole machine thinking has dug out for us.

Treating nature and human society as problems of complication plus Disneyland means giving the wrong answers at every stage, step by step aggravating the situation. What we see is a self-propelling spiral towards disaster. One expression for this – perhaps unnecessarily mechanistic – is 'positive feedback'. Systems governed by positive feedback tend positively in the direction of a basic change of systems, which in human social affairs might mean disintegration. And the last spiral in ACID's historical development had to be the computer – its most refined attempt at machine control: it is so refined that AClD's propaganda manages to convince us that we are dealing with an extension of human thinking, and sensitivity and feeling as well. Even though its intricacy boils down to a simple bipolar scheme (1 or 0, the switch on or off, the house at right angles or not), treating everything as amenable to analysis (the parts tell the whole story – the whole treated as another part), ACID's educators and politicians manage to present it as complex.

A spaceship designed to replace Gaia

ACID's way to deal with the mounting crisis that necessarily results from treating the complex as if it is complicated, is a globalization of the treatment, the thought being that we are faced with a lack of co-ordination: there are still uncontrolled zones that interfere with and disrupt the dataprocessing network. And what can co-ordinate billions of details if not the one and only computer? The perfectly logical far end of this line of reasoning is the global computer, or a globe-spanning network of interlinked computers and 'data-banks', where every computer at any time 'knows' what all the others in the network 'know'. And what they 'know' is the total collection of details about the state of the 'world', mechanistically interpreted into huge crystalline patterns. I call this logical end of the road *Servoglobe*, from 'serve'/'service' and 'servo' in the sense of 'automatically self-adjusting', and 'global'. *Servoglobe* is, then, the historically ultimate stage of the accelerated alienation and one-dimensionality of industrio-competitive society, a stage that means the final abdication of human control and the elimination of (wo)man's creative, meaningful activity as a contributor to Gaia's life. A global, electronic service lady replaces Gaia.[9]

One of the impulses to reflect along these lines came in the early 1970s, in Oslo, when I became acquainted with a man who was employed by one of the leading transnational electronic corporations.[10] This was before the collapse of the Iron Curtain. We seemed to have thoughts and worries in common and one evening after a few beers, he confided in me that the most interesting but also the most worrying part of his work was research on what he called 'the global computer'. The programme is led from an American university, he said, and is supported by several electronics and communications corporations: we have affiliated a large number of researchers, engineers, sociologists, psychologists and language experts, localized in all the major industrial and administrative centres around the world.

The idea is to use satellites for observation and communication, sensor-equipped stations on land and in the oceans as well as local human observers to collect data on the state of the world's various regions. Much research goes into solving the problems of constructing sensible formulations to express the data – including social conditions and political developments – so that the computer can handle them. The world is in deep trouble, he said, and this is the only way to go to overcome those troubles; some are optimists – a few, like myself are worried. Can we do it? And that's only the first part of my worry . . . but what about the Eastern bloc, I queried. They are surely not part of this effort, so how can you globalize without them? They are, by steps, drifting into the same thing, he said. Besides the parties stealing each other's technological secrets, there's a lot more research cooperation than most people can imagine, and because of the sorry state of the world and resources that steadily grow harder to get at, since the ecosphere doesn't respect political territorialities, the scanning and planning systems of the East and West get more and more entwined with each other. And when things break, they break across the boundaries. So in both camps, the global computer slowly appears as the only way to go. The top politicians are hardly aware of what is happening, and among the researchers, the cross-national interlinkage of systems are so far mainly toyed with as interesting science projects. But the development already has so much momentum that it would be very hard to reverse the process. At this stage, when serious ruptures appear, wherever else can they turn, both the East, the West, and the South?[11]

Microelectronics is, after all, the only way they know when complication overwhelms the human mind. But do you know what, he continued, there is one very hopeful aspect to this trend: the Cold War blocs might one day soon wake up and find themselves so involved with each other through this global network that they'll be forced to overcome all their serious differences; they might find their life-or-death prospects are at stake if anything disturbs the completion of this global computer system. That will be a more powerful deterrent than the threat of nuclear war, because it might mean sudden chaos – the loss of the basis for any kind of planning and political manoeuvring, even the use of nuclear deterrent as a workable political tool.

I remember the day well. We were sitting with a good view of the Oslo harbour, where a beautiful square-rigger had just sailed in – its soft, rhythmical lines reminding me of days when qualitative complexity had not yet been replaced by quantitative complication, a time when the human *cybernetes* was – at least in principle – still at the helm; the millennia before Norbert Wiener gave the Greek word another meaning. *Cyberneties* suddenly flashed through my mind as being the great cold turkey knife across human history, replacing Einstein's $E = mc^2$ as the signpost of Western civilization's leap into the abyss. The atom bomb is very visible – the microprocessor creeps upon us, like electrified plastic grass everywhere replacing the real thing. Before it's really noticed there's just the green plastic.

I was not, however, disturbed by the second part of his worry; answering my prodding on that, he said: if the frightful level of complication can be overcome, the computer might take over – we'd have nothing to say anymore! – that, to me, was soap opera science fiction, and it still is, because you can't ever succeed in 'interfacing' the complex and the complicated. *If you have absorbed that deep distinction*, nothing stops that conclusion either.

About this dependency that you are speaking about, I asked, a forced reconciliation of East and West certainly looks like something to be jubilant about; but a second thought comes up: this thing about nothing being permitted to disturb the 'global computer project'. The world being what it is, happily but disturbingly uncoordinated, a lot of wills moving at variance with each other, won't there always be a multitude of threats to that computer-based, globally centralized steering perfection?

Yes, of course – he replied – soft- and hardware going into the network will have to be guarded against any conceivable kind of onslaught. After all, the fate of humankind would be at stake. More would be involved than – say – with radioactive materials deposition. A well-drilled, electronically equipped surveillance police force trained in sociology and psychology, resilient but firm, would have to be constantly on the alert to keep prospective, intentional or non-intentional saboteurs at bay.[12] But in view of the strides that are being taken towards artificial intelligence – what with the competition between Japan and the United States, etc. – we can manage that.

This is how he thought and I've met several computer specialists later looking this way at the world. 'So we'll end up by having a global police state? Is that really something we should hope for as the alternative to the various global and local conflicts?', I went on to ask. Well, at least we would have peace, was his reply, and again, he voiced his own worry, that we'd end up as parts of a global machine, and redundant parts at that. Finally, there would be just that machine.

The perspective was amazing, corning from a serious worker within the field, and not out of a science fiction book. I guess I was gaping at him, and it took me a couple of days before I started thinking that ecophilosophy was immediately relevant to the question, and particularly the concepts 'complication' and 'complexity': 'Life is complex – the machine complicated', and 'the more complicated the machine, the subtler and more difficult to grasp and handle the interface problems – and the greater the consequences of not seeing and handling them properly'. And 'the computer is nothing but a machine, albeit an extremely complicated one'. But 'with a machine like that, you can repair yourself out of trouble for a long while, so that when finally nothing can hold back the system's ruptures, the disintegration accelerates into a chaotic state'.

If this were to happen to the global system the total dependency built up in the meantime would produce worldwide shock and panic. I am not speaking of any normal kind of dependency cut-off, something within the range where history can give us a lesson. What I am talking about is a global socio-ecological crumbling produced by a systematic replacing of complexity with complication on all levels – something unprecedented in mankind's history. That could be the outcome of ACID's last survival convulsions.

Yet we are systematically educated to accept an endless increase in computerization of our world – and our children much more so. There's hardly ever any discussion going on except in a few publicly remote academic books. The task I am proposing is one that should override most others, namely to expose this subtle but gigantic push towards *Servoglobe*, and its consequences.

Notes

1 The 'Gaia theory' was an invention of the British specialist on atmospherical chemistry, James Lovelock, in 1969. It proposes to view the earth's biosphere as one organism and an entity that actively changes itself to survive under shifting circumstances. Since then, a growing interdisciplinary team of scientists has been active in elaborating the theory. ECOROPA – The Campaign for an Ecological Europe – was started in 1977 (this writer was one of the founding members).

2 The ground for this line of thinking had been laid, however, during the years 1955–8, when I worked as an aircraft technician in the Norwegian Air Force. In the middle of that period, we had to be re-schooled to the American 'Technical Order System', under which everything we did had to fit into a pre-arranged, hierarchical order scheme. It gave us an explanation why Vietnamese improvisers won out against the world's mightiest industrial power – at the war's last stage putting its faith in computers. That made things worse, and the jungle guerrillas just kept on eating up the pyramid.

3 A. N. Whitehead's 'Fallacy of Misplaced Concreteness' is related to this.

4 The ability to improvise, individually and collectively, is now (1999, 2000) being studied at the University of Trondheim, Norway, through the project 'Improvisation in Interdisciplinary Light'. Various cultures are here being compared. Metaphors, enlightening to the problematique, are the jazzband and the European symphony orchestra, contrasted.

5 Schumacher explains that he got this set of ideas from a study of Burmese Buddhist culture.

6 Zbigniev Brzezinski: *Between Two Ages: America's Role in the Technetronic Era*, New York 1970, Greenwood 1987. Brzezinski was a special adviser to President Jimmy Carter.

7 The difference between the Chinese and the Indian branches of Buddhism, where the latter was dynamically oriented, was reflected already at the famous Council of Lhasa

1,000 years ago, where the Chinese were thrown out on the ground of their static interpretation!

8 In 1985 I visited the Artificial Intelligence laboratory at the University of Texas, Austin, and interviewed its director. Philosophically still in the seventeenth century, he was convinced that at the next step computers would feel and value, the interface with living entities being near to perfection. It was just a matter of miniaturization and parallel processing. My quoting Hegel, where he says that he who lacks a knowledge of the history of philosophy is doomed to repeat it, just produced a blank stare; any discussion of basic presuppositions did not have a chance.

9 Since I wrote the first version of this, a British computer expert, Geoff Simons, has come out with a book, *Eco-Computer*, describing in detail how the global computer (*Servoglobe*) is becoming a reality (New York: 1987).

10 Xerox International.

11 India has by now, 1999, one of the world's largest communities of IT developers, probably third after the United States and Japan. But they are pushing in the same direction; no Indian philosophy (Hindu or Buddhist) is involved. Take note, though, so far, only 1 per cent of mankind has Internet access!

12 One aspect of this, the attacks by 'hackers', has grown strong since the original version of this chapter was written. Their threat to the Internet function (and to *Servoglobe*) is now met by surveillance systems that 'could compromise the anonymity and privacy of law-abiding citizens' – in other words, 'Big Brother' (Orwell). The vulnerability of the globalized 'infomation highway' produces one more spiral towards *Servoglobe*. See *New Scientist*, 19 February 2000, p. 6.

Bibliography

Literature that has inspired my writing (partly referred to above), and that I would like to recommend as elaboration and substantiation of what I have said:

Burnham, David: *The Rise of the Computer State*, London: 1983.

Dreyfus, H.L. and Dreyfus, S.E.: *Mind over Machine*, New York: 1988.

Ellul, Jaques: *The Technological Society*, New York: 1965 (original edition: *La Technique*, 1954).

Gregeseu-Roegen, Nicholas: *The Entropy Law and the Economic Process*, Cambridge, Massachusetts: 1971.

Haugland, John: *Artificial Intelligence – The Very Idea*, Cambridge, Massachusetts/London: 1985.

Hodder-Williams, Christopher: *A Fistful of Digits*, London: 1968.

Koestler, A. and Smythies, J.R. (editors): *Beyond Reductionism – New Perspectives in the Life Sciences*, London: 1969.

Lovelock, J.: *Gaia – a New Look at Life on Earth*, Oxford: 1987

Mumford, Lewis: *The Myth of the Machine*, New York: vol. I 1967, vol.II 1970.

Roszak, Theodore: *The Cult of Information – The Folklore of Computers and the True Art of Thinking*, New York: 1986.

Schumacher, E.F.: *Small is Beautiful*, London: 1974.

Searle, John: *Minds, Brains and Science – The 1984 Reith Lectures*, London/New York: 1989.

Shallis, Michael: *The Silicon Idol – The Micro Revolution and its Social Implications*, Oxford/Melbourne: 1985.

Waddington, C.H.: *Tools for Thought*, St. Albans: 1977.

Weizenbaum, Joseph: *Computer Power and Human Reason – From Judgement to Calculation*, London/New York: 1984.

Whitehead, A.N.: *Science and the Modern World* (Lowell Lectures 1925), New York: 1949.

Some of my own writings arguing or developing relevant items:

'Kraft-rom mot elve-tid' (Power-Space versus River-Time), in: Haagensen, K. and Midttun, A. (editors): *Kraftutbygging, konflikt og aksjoner* (Hydroelectric Development, Conflict and Actions), Oslo: 1984.

'Mangfold og tid. Hvordan en bryter pyramidens herredømme' (Diversity and Time. How One Breaks the Reign of the Pyramid), in: Skönberg, S. (editor): *Grønn pepper i turbinene* (Green Pepper in the Turbines), Oslo: 1985.

'Økofilosofi versus New Age' (Ecophilosophy versus New Age), in: Gule, L., and Laugerud, H. (editors): *Vitenskap og verdensbilder* (Science and World Views), Bergen: 1989.

'Servoglobus – Den siste pyramiden' (Servoglobe – The Last Pyramid), in *Kontrast,* no. 122, 3–4, Oslo: 1990.

'Økokrisefilosofi – glimt fra det norske økofilosofiske forsøket' (Ecocrisis Philosophy – Glimpses of the Norwegian Attempt at an Ecophilosophy) in: Gjerdåker, S., Gule, L., and Hagtvedt, B. (editors): *Den uoverstigelige grense – Tanke og handling i miljøkampen* (The Unsupassable Limit, Thought and Action in the Environmental Battle), Oslo/Bergen: 1991.

'Inside Nature', in: Goldsmith, E. *et al.* (editors): *The Future of Progress*, Bristol/Berkeley: 1992.

'Systemkrise og krisesystem' (Systems Crisis and Crisis Systems): in: Album, G., Alstadheim, K.B., and Andersen, G. (editors): *Supermarked eller felles framtid? – EF, økokrisen og Norges valg* (Super-market or Common Future? – European Union, Ecocrisis and Norway's Choice), Oslo: 1991.

Naturens nei (Nature's No), Norsk Bonde-og Småbrukarlag (Norway's Small Farmers Association), Oslo: 1994.

'Mother Earth's Treasures and their Revealers – An Ecophilosophical Perspective', in: Kuløy, H.K. (editor): *The Treasure Revealer of Bhutan*, Bibliotheca Himalayica, Series III, Volume 8, Kathmandu: 1995.

8

A PROVOCATIVE PESSIMISM

A postscript on the scientific worldview and global order

Georg Henrik von Wright

(THIS TEXT ORIGINALLY APPEARED IN SWEDISH IN 1993 AS A POSTSCRIPT OF THE BOOK *MYTENOMFRAMSTEGET* (*THE MYTH OF PROGRESS*), PUBLISHED IN STOCKHOLM BY BONNIERS)

> It is beginning to dawn on us that our erring ways
> may have even greater consequences than we first surmised.
>
> Harry Martinson, *Aniara* 13

It's been six years since the publication of *Science and Reason*. Much of what I have written in the meantime has been an attempt to work out some of the themes pursued in that book. In this chapter I shall stop for a minute to confront the ideas in my book with what I would call the current situation.

The book had a subtitle, *An Attempted Orientation*. Actually, the attempt went in two directions. Partly, I wanted to clarify for my own sake the main attributes of what I call the scientific worldview – its relations to earlier cultures and their efforts to understand reality rationally, and how the worldview has changed since its emergence in the late renaissance and early baroque up to our own time. Partly I wanted to try and assess the impact of science in terms of technology and industrialization on life.

I would like to mention in passing that I use the word 'science' in its English sense – that is, mathematics, the physical sciences and other exact empirical studies – and that I mean by technology an ability to control and intervene in processes in nature, an ability resting on scientific knowledge. The humanities, then, are not included in my concept of science; nor will I discuss pre-scientific mechanics as an expression of technology.

So, in keeping this sketch of the book's purpose, to confront its thinking with the current situation entails two things:

The first is science. The question here is whether we can observe results or trends in science in recent years which could be said to have added something fundamentally new to the scientific worldview.

The second question is about the current global situation. Would the crisis in mankind's relations with the physical and social environment – caused by the impact of science-based technology on our existential state – be exacerbated if viewed from a global perspective?

First, then, a few reflections on the worldview.

The modern scientific worldview came into being in the early modern period and attained what I will call its classic Gestalt in the three centuries between Galileo and Einstein.

One of the essential conceptual assumptions of this worldview – we could call it a claim or postulate – is the idea that all events in nature are strictly governed by law. It follows, then, that all – at least all the important – phenomena can be explained as the outcome of rule-bound causes. If such causes occur repeatedly, it should be possible to predict the outcome: if the causes can be reproduced, then the effects can be produced artificially, or suppressed. More or less everthing called technology is indebted to this realization. Therefore, what we know of the natural laws gives us, ultimately, a means to govern and control what happens in nature.

A worldview built around the concepts of law-boundedness, explanation, predictability and control could be called causally deterministic. This is, or more accurately was, the classic Western scientific worldview. It was coherent, uniform and imparted therefore a feeling that reality was comprehensible – if not in every detail, at least in principle. It also cohered with the relative stability of the cultural climate – and, on that account, a conception of the cultural of the Western form of civilization.

In the early years of the twentieth century, natural science was shaken by profound internal mutations; indeed they threatened to undermine the very basis of the classical worldview – with principal reference to causality in nature. The most important of these changes had their origin in quantum mechanics or quantum theory. Heisenberg's uncertainty principle was particularly significant. The principle states that it is impossible to know the exact position and velocity of certain minute particles at the same moment in time. It set a limit on measurability which, in turn, is the basic principle on which exact prediction rests. Chance had intruded in the worldview – something conservatively minded scientists were loathe to accept. It was this state of affairs that prompted Einstein to coin his oft-quoted phrase, 'God doesn't play dice'.

Heisenberg made his discovery in 1925. Only a few years later, a genius of a mathematician, Kurt Gödel, showed that in a sufficiently rich mathematical system, there are intuitively truthful postulates which in principle cannot be settled, that is, confirmed or falsified within the system. So not only is physical measurability hemmed in at all sides by barriers over which the inquiring mind cannot spring, logico-mathematical provability is too.

These new insights can be said to have dealt the classic worldview of European science a crippling body blow as early as the beginning of our century. There was talk of a *Grundlagenkrisis* – a crack in the bedrock of physics and mathematics. Old models of nature's composition and laws, and the power of the human mind, were brought into question. New models or logico-mathematical programmes of the same overarching character have so far been unable to replace the old. Nor would it be wrong to say that a scientific worldview no longer exists with something of the same coherence, comprehensibility, intelligibility, as that of the past two centuries, especially that of the late nineteenth century. The worldview is slowly coming apart. And its role as a cultural mainstay is necessarily changing as well.

I discussed briefly in my book the emergence of new scientific approaches usually called holistic, in contrast to traditional mechanistic or meristic explanatory models. These new ways of thinking are particularly prevalent in the study of biology. But their influence is also apparent in physics and chemistry. Nevertheless, I shall not discuss holism here; I have to admit, an intelligible idea of what it means I have yet to encounter. Instead, I want to ponder other events in contemporary science which, over the past few decades – one could even say over the past few years – are of such gravity, they have arguably changed, or at least begun to change, the worldview itself. What I have in mind are tendencies partly in mathematics, partly in physics/chemistry, an area to which one tends to refer by the admittedly not entirely inappropriate name of *chaos studies*.

A leading figure in the field is Ilya Prigogine, Nobel laureate in chemistry. Another leading chaos theoretician is the physicist Herman Haken, a third mathematician René Thom, whose most important contribution goes under the rubric of catastrophe theory. Prigogine and Haken have developed, among other things, models for dynamic systems whose energy state interacts with their environment. Unlike closed systems, which grow increasingly disordered, the so-called heat death at which all energy differences are extinguished, the new order-structures originate in the dynamic systems, instances of which are what we call living phenomena. They survive by 'sucking order from their environment', as some would have it. They consume what in thermodynamic jargon is called negentropy or exergy, ultimately helping to speed up the process of entropy, or disorder, in the universe as a whole.

The aspect of the new research which possibly more than anything explains why it is called the study of chaos, has to do with events where infinitesimal, imperceptible, differences in an initial state can result in highly different states which are impossible to predict beforehand. Reference is often made to the butterfly effect. The flapping of a butterfly's wings somewhere in the Caribbean can be the ultimate cause of a North Sea tempest. An effect no one observing the flitting butterfly wings could have predicted.

Another side of the same variety of research is about systems of a holistic nature. Ecosystems are good examples. Minor disturbances, such as the extinction of a biological species or introduction of an alien species, may have unforeseen consequences for the system's equilibrium. There are several striking examples of this, but in terms of theory, the field remains rather undeveloped.

We can discuss whether the unpredictability of the processes is due to causelessness in reality or simply our limited cognitive abilities – whether the limits are ontic or epistemic, to borrow from the philosophical vernacular. A similar discussion took place – and is still going on to some extent – around the time of Heisenberg's uncertainty principle. Whether God does in fact play dice with the universe or not, then, still awaits an answer!

But whatever the answer to the *ontic–epistemic* question, the fact remains that our notions of the predictability of natural processes are changing. And in the process, the classic scientific worldview is losing some of its, if I may say it, deterministic nimbus.

Now, one might ask what the general cultural repercussions of these changes in the scientific worldview are having, or might arguably have. That science exerts an effect on culture and social life from the outside is indisputable. It is true not least about what I have called the classic worldview.

One can always instance the enormous significance of Newton's mechanics, especially in the latter part of the eighteenth century, as a model, or to use a popular term today, a paradigm for thinking about a society, economy, upbringing and absolutely all forms of human or inter-state relations. One sought in these relations too – in the same way as with the movement of bodies – strict, deterministic law-boundedness; the trick was simply to discover and describe the determinants and outcome constellations in the right way. Put somewhat bluntly, we could say that it was an art which we, since those days, have been trying to practise in the social sciences with varying, but hardly great, success.

To return to chaos studies – if I might use the slightly unflattering epithet – they teach us to have respect for the unpredictable, and, in turn perhaps, have fewer expectations of our ability to systematically predict and shape our societies' future.

Long-term planning has been a central issue in most nations' social policy in the past decades – whether they subscribe to a market, planned or mixed economy. Planning has acquired a scientific substratum of various theories on strategic thinking and cybernetic control mechanisms, theories which largely emerged in the second half of the present century, inspired not least by progress in scientific combat technology during the Second World War.

Social planning appears now to be in a crisis, however. In many cases it has shown itself to be what one calls in English *counter-productive*, it has worked against itself, and at exactly the same time as it has shown itself to have undesirable and unforeseeable consequences. The optimism nourished by the extraordinary and spectacular advances in optimization and programming mathematics in the first post-war decades is taking a doubtless necessary lesson from the scepticism which in turn is fed on mathematical catastrophe theory and chaos studies.

The unexpected and confusing political events of recent years have also fostered further insight into the limits of social planning. The total collapse of the Soviet empire, Germany's re-unification, the advance of religious fundamentalism as a force in global politics, just to mention a few.

These external changes were not precipitated, of course, by chaos studies. Nor have they affected the scientific theories on chaotic events and predictability.

Nonetheless, their more or less concurrent manifestation is hardly a coincidence. In as turbulent a time as ours, terms like equilibrium, instability, predictability insinuate themselves into the collective conscious almost by necessity. They colour, moreover, how we understand our efforts to master events in the world we live in. But how it will all turn out is not easy to say; these changes are still in their infancy.

So much for the worldview – now to something about the world itself.

We have already mentioned how organizational and technical progress facilitated by science has also called into question mankind's relations with nature, that is, the reality mankind manipulates with its knowledge.

We see today – more clearly, I suggest, than even five or six years ago – that the question has two sides to it. I shall call the first ecological and the second ethical. The ecological aspect concerns for the most part what we call environmental and resource-related issues.

First, a few words on how I tell the two aspects apart. Of course we need to remember that while they are distinct, they are also intimately entwined.

Typical environmental questions are what we could also call technical problems. As such they are – at least frequently and largely – value-neutral. There is a given objective, such as to reduce the environmental damage caused by a given industry. To achieve this end, we ask the advice of experts with the necessary scientific and technical qualifications. Whether the proposed means for achieving the end work or not can be ascertained objectively, in principle, and will be something about which everyone can agree.

Ethical problems are different. They are about both means and ends. Which objectives are legitimate? Which mechanisms do we have a right to use to reach the objective?

These questions are not value-neutral. They cannot, in a certain sense, be determined objectively. Which does not put us at the mercy of preferences or discretionary judgements. Through argument and discussion, we can agree on a way forward. But this requires a community of values in some shape or form, to which the implicated parties subscribe.

The ethical dimension of relations between humankind and nature is primarily about our relations with the nature we – and kindred creatures – represent through our bodies. The problem is acute given the amazing advances in medical technology and genetics. What characterizes the current ethical debate in science are the prospects opened by a technology *in statu nascendi*.

Related to the ethical debate prompted by progress in medicine and genetics, there is the question of what are called animal rights, in connection, for instance, with transplant surgery or animal breeding for industrial food production or manufacture of other products.

An ethical question of a slightly different kind concerns our responsibility for coming generations. Do we have a right to exploit natural resources and plunder nature in ways which endanger or simply make it impossible for our descendents to lead a life of the type we ourselves consider justifiable and humane?

Ethical views can naturally also be applied to our dealings with nature as a whole. Do different types of landscape have an intrinsic value requiring our respect? Or the multitude of plant and animal species? We can say that absolutely every long-term policy in the field of environmental conservation and preservation has an ethical dimension which is likely to foment value conflicts.

I shall nevertheless not pursue the ethical aspects of our theme here. I honestly admit that I have found it, and still find it, extremely difficult to get a firm grip on them. It may have something to do with the subjectivistic and relativistic points of departure of my own philosophical peregrinations in ethics. This is a subjective judgement on my own insufficiency. But I also believe there is an objective reason for my difficulties which might be worth a moment's attention.

It is a well-known and often-attested fact that the community of values which formerly bound us Western humans together, built largely on our common Christian cultural and religious heritage, is slowly eroding, vanishing, dissolving. It is being supplanted by something we could call an empty space, or value vacuum, where commonly accepted benchmarks against which we weighed and assessed ethical standpoints have lost their hold.

How we shall extricate ourselves from this state of moral dissolution, I'm sure I don't know. No one knows – even though some believe they have the answer. Some put their faith in a religious renaissance. I find this difficult, myself. The fundamentalist leanings which are today so conspicuous a part of contemporary religious life, fill me with horror and dread. They are not only flourishing in the Muslim part of the world. Similar tendencies are observed in what used to be Christianity's eastern half. And, not least, in different superstitious cults in the West, especially perhaps in the United States. They manifest themselves as astrology, creationism and other quasi-scientific humbug.

Another form of fundamentalism, in its way, is the mindset usually called scientism, which consists of uncritical and exaggerated belief in the ability of techno-scientific, targeted rationality to solve all of our problems; such questions as fall beyond its competence can only be the subject of opinions, not reasonable deliberation.

What I have to offer to the debate on science and ethics is not, in other words, much. But what I believe I have made clear about the value-nihilism of our age, is, nevertheless, important to note if we want to conduct a rational discussion on salient questions at all. We risk otherwise participating in a discussion from which emerge prosaic recommendations or pious aspirations.

I shall now turn to the ecological aspects of our subject matter. For my part, I find them – possibly in contrast to the ethical – highly substantive, not to mention urgent. Awareness of them has grown sharply in the few years since *Science and Reason* came out.

It is no longer a question of air and water pollution, desertification and forest decline. Man-made, possibly irreversible natural processes like the thinning of the ozone layer in the stratosphere and global warming, are today threatening the survival of life on earth. Five or six years ago, we didn't hear much about them.

Another difference is that the warnings are no longer given by the environmental activists and scientists alone. The political powers that be and representatives of the economic, industrial and technological establishment are also nervous. Even the most reactionary and uninformed popular leaders in commerce and industry are going around in green anoraks.

We can call it the awakening of a global ecological conscience. It occasionally expresses itself in what I would call environmental hysteria. Scientists and politicians are calling in unison for something radical to be done. Now something is being done, and increasingly – but still, I would suggest, very, very little, also in relation to the desired goals which enjoy the general backing of the public. Knowledge and action are, at least for the time being, on opposite, mutually antagonistic sides.

One important reason for this state of paralysis lies, in my opinion, in the inherent contradictions in perceptions of the ideal we should be aiming to realize. An assertion I shall attempt to justify a little more thoroughly.

The Brundtland Report highlighted the concept of sustainable development, an already well-known concept moreover. It is discussed in the report in some detail and has become something of a slogan. But it remains, at least for me, hazy and difficult to fathom. The picture emerging on the basis of the report and ensuing discussions looks more or less like this:

It is imperative to minimize the depletion of natural resources and destruction of the environment. The downright frightening devastation of the environment observed in different parts of the collapsed Soviet empire – Central Asia, Kola Peninsula, St Petersburg region, including north-west Estonia, but also parts of Poland, Czechoslovakia and former DDR – must not only be halted, but reversed if at all possible. The affluent countries' mania for consumption must end, energy must be saved, and pointless consumption contained. Decisions on the latter are necessary, not only for the sake of the future of the industrial nations, but also to save the poor parts of the world from sinking further into misery, a situation which grows worse by the day relative to that of the wealthy countries, but also in many areas absolutely.

This on the one side. But on the other, growth in the industrialized countries should continue, living standards should not be allowed to fall, if possible; on the contrary, they should rise; and people should not be forced to get by without the new toys high-tech industries disgorge at an increasing rate into an ever-widening geographical market.

Attempts are made to explain how these two opposing tendencies or efforts depend on rather than contradict each other. Without growth, it will be impossible to redistribute wealth more equitably. 'We have to get richer first, to help the poor later.' The underlying idea in these and similar arguments is not hard to follow – and they are not entirely devoid of merit either.

Another argument in favour of growth sees the root of the evil in the population explosion. The misery of the world is equal largely to the hair-raising poverty of the most populous nations. The cure for poverty is economic growth. With rising standards of living, population growth will slow to a halt – this is what the situation in developing countries shows. So we need to accelerate growth if we want to save the earth.

The short-sightedness of the argument is its tacit assumption that there is time to spare for this self-correcting mode of population control to work. It is an assumption, I have to admit, that I find to be highly untenable. At the very least it should be debated, not tacitly condoned. If the poor – roughly four-fifths of the human race – among which the peoples of the former Soviet Union should be counted – are to realize the same standard of living as the average inhabitant of the affluent part of the world, with everything that implies in terms of energy and material consumption, besides what is left over in the form of waste, the earth would probably be taxed more than it could bear. Long before the poor enjoyed the wealth required for voluntary birth control to take hold, ecological and demographical disasters would have shaken the world to the core. The patient might even die on the way to what is considered a 'natural recovery'.

I know of not a single convincing or even serious argument for the possibility of continual growth. The arguments I have heard I have found to be either completely groundless or also escapist fantasy – like major mining operations on the moon, or relocating a significant proportion of the earth's population to artificial satellites in space, or to the planet Mars, which has been made habitable. Myself, I consider these perspectives far less likely than the idea that humanity, through its colonization and exploitation of the earth, has demonstrated its incompetence as a biological species, and therefore condemned itself to extinction. I have been accused of doom-mongering. But I believe that if we could only learn to take the measured view, we could make a serious effort to ensure that it doesn't come to pass.

My final proposition must therefore be that the contradiction I believe is intrinsic to the concept of sustainable development can only be resolved by a movement which is the opposite of the current expansive, growth promoting tendency. It is not only about containing the population explosion. We need something that could be called a downsizing or cooling down of the world's economy per se.

A development of the type I am here advocating, would require people in the industrialized world to change their lifestyles, put up with a much more frugal life than today, and relinquish desires they are used to satisfying, but from which we can abstain without coming to harm. We in the industrialized countries must quite simply become poorer, on average, to avoid exacerbating the polarity between developed and developing countries, and also between A and B citizens in the former, and destroying the balance of the planet entirely.

I believe many agree with me about the direction developments should take. Its desirability, even necessity, has often been highlighted. It is not particularly novel, rather trivial in fact, to insist that continued economic growth will at the end of the day be physically impossible. It is far more complicated, but no less important, to envisage how it could be accomplished, and whether it is even possible.

How could it be done? A resetting of the balance between production and consumption is unlikely to happen of its own accord. It would seem to require some sort of regulatory control of what are usually called market forces. The regulatory authority could be the state. The state as a regulating hand in the economy does not mean socialism in the conventional sense, or that the state acquires ownership of the means of production. It does require, however, the state to have political authority over the economic system – that the state can articulate a will and force it through, coercively if necessary. This authority or sovereignty has also traditionally been an attribute of state formations known as nation-states and which in the West are almost without exception constituted as parliamentary democracies. This state form is on the decline, or perhaps disintegrating is more apt, crumbling little by little.

The ongoing process of integration in Europe is a sign of this type of decline or collapse. The former nation-states seem to have coalesced into a new form of union, where the economic system is superior to the political, or takes on some of the political system's former role. This is a remarkable development. The economic, 'the market', becomes in so many words the sovereign power, or, as we might feel tempted to say in light of its apparently unstoppable urge to expand: the economy has bolted. The space vacated by the downfall of the planned economy has been filled by galloping capitalism which seems to be dragging virtually every nation of the world and their governing bodies in tow.

Put bluntly, we can therefore say that the thought model on which the European Community is built, is a reinforcement and extension of precisely the production and consumption patterns that have made global balance such a problematic issue. The European Community is by nature a system of economic growth, expressing its nature internally by dissolving former trade barriers between member countries – and externally by being an equal competitor to the other two economic giants, Japan and the United States. In human terms, the Soviet empire's exit from the political area creates untold opportunities for economic expansion in Europe's eastern enclaves.

Now the EU has adopted rules to regulate the economic behaviour of its member states, but these rules are not of the type that slow or control growth. Their purpose is, on the contrary, to protect market forces from competition distorting factors. To these belong state monopolies and government preferred industries – including subsidies for agriculture and other industries. The counterforce intended to guarantee market competition is called privatization. It undermines the state's economic clout and threatens its ability to sustain a welfare society which the European democracies struggled to put in place, and where the Nordic countries, Sweden first and foremost, have been the trailblazers. The Swedish model has lost much of its allure.

Hence, Europe's ongoing process of integration can be said to be moving in the opposite direction to what I would consider necessary – if humanity's journey into the future is not to come to a disastrous conclusion. I have not intended to say in expressing my qualms about the EU's innate economic expansionism that trade barriers in the shape of custom duties and import restrictions or state monopolies or state subventions are the way to go to facilitate the slimming of the economy I believe must take place.

For that route I neither can nor will provide a tour guide. Taking that route is not dependent on policy decisions on the economy primarily, but on a change in our attitudes, in our conception of what human beings need, what constitutes a good life, and acquiring a sense of global solidarity which does not protect and place our own material standard at the centre of recipes telling all the others how they should live. We can piously hope for our mindset to change, and we can also agree on its necessity as long as we can avert disasters of global proportions which at the end of the day will affect every one of us. But it is a long way to go from wishes and insights to changing the real forces behind economic and political developments today and their defence by security forces.

Is it not inconceivable that the market economy's capacity for self-regulation protects it from devouring its own? A regulatory device not coercively imposed by external laws and regulations, but by the hard, practical necessity of saving one's own skin. One way in which this could happen is if the stocks of raw materials for industry dried up or became inaccessible and the markets diminished or disappeared. It is a brutal perspective – associated with widespread human suffering. Because it means, without beating about the bush, that the growth spiral will stop only when it runs out of its own steam.

I believe it is important to let these ideas take us wherever they will. And they lead us into an impenetrable darkness. From this vantage point, building a bridge from a state which *is* to what *ought to be* seems like a political and – at least for the time being – also a popular psychological Utopia without basis in reality.

With these contradictions and impossibilities in mind, I have found it increasingly difficult to believe in a sensible and systematic resolution of the global dilemma. The sense of pessimism expressed in *Science and Reason* has deepened over the years. Unlike my statement at the end of that book, I am unable to confess to any faith in reason as a hope for humanity.

In these circumstances, it is tempting of course to shrug one's shoulders and let everything carry on as always. I suspect, however, that resigning in the face of the apparently unavoidable has already happened in most people's hearts, even if their tongues do continue to profess hope. We will be hearing these confessions for a long time to come, in the form of resolutions from international conferences and symposia, committee reports, agreements on reducing exhaust and other emissions, and attempts to reconstitute an already devastated nature.

While at the same time negotiations proceed around other tables on wage increases and tax cuts, and on which economy stimulating steps are likely, as the expression has it, to strengthen the business's international competitiveness. As long as a growth philosophy remains the predominant ideology, the inextricable purpose, whatever is decided at the latter tables will make wishes expressed at the former more difficult or simply impossible to realize in any foreseeable future.

The author of *Science and Reason* has been repeatedly taken to task for expressing a sense of pessimism likely to create anxiety and paralysis.

And to a certain extent, it is not an incorrect description of the situation. But I find the optimism whereby all one needs to do is to sit back and let things work themselves out knowing that more research, new technologies and the uninhibited play of market forces will put everything together again far more reckless and paralysing. This, I believe, is the optimism of the pit of powerless into which governments have descended, from where they attempt to anaesthetize the masses over which they govern.

We must fight such false optimism with the tools at our disposal. But I don't think we will be rid of it before experiencing the desperation which a sober assessment of the human situation in my opinion will force us to acknowledge.

One could perhaps call my approach a provocative pessimism. I came across the expression in the writings of another person for whom the situation has also been a subject of a highly critical appraisal, Hartvig Sætra.

One idea to have found expression in the poetry of many civilizations and religious mythologies is that humanity can only attain the wisdom which changes its way of life through suffering and ordeal. These ordeals can affect people in many different ways. In the life of the common people, they can be caused by famine, war, pestilence and great natural disasters. It is possible that we are standing on the brink of an age in which disasters will strike hard and pitilessly. Whether they will make us any the wiser in the longer term is debatable. On the other hand, these events could affect our conditions of life and, thereby, our way of living it.

The global order is looking increasingly like a system of the type students of chaos investigate with bifurcation points from which developments leap forward in completely unpredicted and unforeseen directions. It creates a bigger space for the irrational. This is a fact which the rational observer must recognize and take into account. It doesn't mean that we shall respect the irrational. But it could, perhaps, teach us an intellectual lesson in humility which could come in useful, not only for our theoretical understanding but the practical planning of our own future. Much of what has made our existence problematical stems from overestimating our capacity to rationally master, with the aid of science and technology, the situation.

9

GLOBAL AND LOCAL

Torsten Hägerstrand

(PUBLISHED IN THE SWEDISH GEOGRAPHICAL YEARBOOK OF 1989)

An outline of the problems

The list of global problems that are widely recognized can be made very long. Many countries are approaching a population meltdown. Only 8 per cent of the global population live in countries with zero population growth. Starvation and malnutrition are keeping on. The greenhouse effect will reportedly change the climate and raise sea levels. The ozone layer is threatened. The volume of waste is multiplying and there is no systematic way of dealing with hazardous substances. Coastal waters are polluted in many places, and groundwater looks next in line. The tropical forests are laid waste and farming land eroded. Plant and animal species are becoming extinct. The global supply of energy has not been solved satisfactorily. The state of planet's life-sustaining cover tells us that it's high time to impose global limits on extraction and ensure respect for them going forward.

SCOPE (The Scientific Committee on Problems of the Environment), in its mission statement on the study of environmental problems 1989–1991, urges the scientific community to provide the critical information necessary for ensuring the environment is taken into consideration when national and international strategic development plans are laid. Hearing these exhortations and knowing the complexity of the problems, one wonders what sort of contingency plans are in place in scientific and political quarters. This is the question I shall be addressing in the following.

When confronted by really serious problems, we often look for levers of sufficient length and strength to put things onto a new, more acceptable course. As far as anthropogenic global problems are concerned, UN resolutions, international treaties and national law-making would seem to offer the levers we need.

Such measures may not be effective enough, though, even if they are necessary first step. Trials with a top-down management style to control growth in the

Third World have not resulted in a convincing model. There is a sensible review in the *Journal of Economic Geography* (1988) of the book *Sustainable Development of the Biosphere* which touches on the issue. Highlighting the importance of a global perspective, the reviewer, Vaclav Smil, concludes with the following comment:

> [t]he evident existence of global problems does not gainsay the essential principle of an effective management: to act as locally as possible. This is how a successful reforestation project worked in South Korea; it is how the Japanese manufacture excellent automobiles (small quality groups in large assembly halls); it is how shelterbelts and no-till ploughing keeps erosion at bay in Dakota. The particular features of a local management system obviously can't be transposed from one place to another. The tension between large-scale generalization and the uniqueness of different areas is the essential thing in geographical studies. It is why geographers should welcome the now popular global studies and yet mistrust their practical effect.

His remarks bear on an epistemological problem which in general has been insufficiently deliberated, and that is the mutual interaction of large- and small-scale phenomena and processes. Eddington, one of the great interpreters of relativity theory, said in a speech on the subject (1935), 'Had our eyes been of different sizes, we might have developed an ability to combine the perspectives of the mammoth and the microbe'. But they aren't, and in any case the eyes of a mammoth aren't big enough to see the contexts which need to be seen.

The landscape mantle

Conceptually, we are talking about the global and the local as if they were opposite poles with nothing in between them. In fact, we are dealing with two different approaches to what Friedrich Ratzel described as 'that thing-filled surface of the earth'. *Thing* should be taken in its widest sense. We do not differentiate between inanimate and animate things or natural and artificial things. The human mind, with its knowledge and intentionality, is certainly not a physical thing, but since it inhabits each and every one of us, it has a thing-like presence in space and time. And for that reason it can be counted among the things filling up the surface of the planet.

Looking at it as from the perspective of a moment in time, Ratzel appears to have been thinking of a random collection of unrelated things. But if we add the temporal dimension, a web of intricate relations emerges. Since we need an agreeable term for the purpose, we can call this web of things and relations the earth's *landscape mantle*.[1] The term geobiosphere wouldn't suffice. It excludes human institutions and technology. And that wouldn't be realistic.

The landscape mantle is characterized by two groups of ordered principles. The first of them contains the principles of nature set down through geophysical and biological evolution. Already here, a plethora of links connect the local and global.

The major links are the circulation of air and water. The weather and precipitation vary locally, but the climate is driven by global circulations in league with the major ocean currents. New species of plants and animals split off from their predecessors locally before dispersing more or less widely over the surface of the earth as vacant habitats fell within the reach of the dispersal mechanisms.

The global significance of the biogeochemical cycle involving water surfaces, organic life and atmosphere is a relatively recent discovery. The comprehensive international geobiosphere programme (called *Global Change*) seeks above all to determine the character and extent of these currents.

The second group of ordered principles is man-made – if we can speak of principles in this sense. There are similarities between them and certain natural processes. So, for example, new ideas and discoveries occur locally just as new organisms and disperse more or less widely either spontaneously by imitation or driven by marketing. Technologies need habitats just like organisms.

Our reliance on nature's food chain is essentially no different from that of other higher animals. On the other hand, though, economic institutions and physical means of transport have together given urban and industrial populations an opportunity to enjoy non-locally produced goods. As a result, the links are less apparent than when transportation was less developed.

A cultural feature of incomparable importance in the landscape mantle is the partitioning of the land – and, to a degree, the oceans – into domains. National territories are at least for the time being of the highest political significance. The purely physical changes in the landscape mantle are, conversely, mostly tied to details in land ownership or usufructary rights. Property could, if you like, be seen as advanced versions of animal's territories. The purpose is similar: to protect vital supplies from unwanted invasion. The difference is that property is part of a hierarchy of domains. It is the lowest rung of the system by which national governing bodies seek to control what may and what may not be done, or cannot be deviated from, with regard to the use of land, water and air. The domain system comes naturally in many different shapes and sizes. The number of levels varies with the size of the country and administrative traditions. In large swathes of the developing world, access to land is allocated according to custom without perceptible borders on the ground. Sector bodies can cut across the geographically delimited domains. The property rights themselves, by the way, can be heterogeneous. There are examples, as in some oases in African countries, where a field, the water and palm trees on the same plot of land are owned by different individuals (Eldblom 1968).

It remains to mention the cultural element, which today links the local and global far more effectively than ever before in our history. The electrically powered infrastructure, enabling the speedy exchange of commodities among actors the world over, constitutes a unique expansion of interpersonal verbal contact. There is nothing like it in nature. Modern communications technology facilitates participation of local communities and enterprises in the global economy. The mass media have at the same time evolved into a mental 'climate factor'.

The tyranny of the small decisions

Given the global problems exemplified above, how can we explain our conduct? Science has taken giant strides during the period and technology is at the same time increasingly vital to scientific discoveries. On the one hand then our intellectual accomplishments are of the highest order, but on the other, the result is more than doubtful.

A discussion of the role of science and technology must wait while something is said about the conditions of human action. We are always trapped between what appears to be a meaningful objective and our – individual or collective – physical capacity to bring it to fruition. As soon as we satisfy our basic needs for food, clothing and shelter, there is no real reason why we should feel an urge to add incessantly to our material possessions, and in an escalating tempo. But judging from the behaviour of the privileged classes in times past and during periods of our own history, this is what we have chosen to do whenever the opportunity presented itself. The dramatically new is that thanks to the industrial system the whole population can have a share in the products of unprecedented material growth.

The individual's ability to consume natural resources or produce emissions and waste is, however, not particularly significant. From a global perspective, the impact of the major corporations on the environment taken separately is not all that noticeable. It's the quantity that matters, of course. Much of what in isolation makes sense and seems innocent enough cannot be multiplied beyond a certain point without major social or environmental problems – or both. The landscape mantle isn't wide enough to absorb everyone's desires.

The dilemma facing human society is the cumulative outcome of 'the tyranny of the small decisions' (Odum 1989). Completely innocent micro actions have had unforeseen macro effects through duplication and accelerating dispersal. The art of multiplication has not been employed to estimate the impact and exercise control of individual choices. As an example, Odum points to how people everywhere got rid of the problem of industrial smog by increasing the height of chimneys or flue gas stacks, as if magicking the odious discharges away whenever the wind managed to blow the fumes out of eyeshot.

He could have mentioned the car as another example. Its continuing dispersal is another troubling cumulative process even if we succeed in removing current exhaust gases. The mobility provided by the car cannot be enjoyed by every person in the world without an unacceptable price in the form of loss of land and energy.

It shouldn't come as a surprise, really, that local actions are never assessed in light of cumulative consequences beyond the local horizon. Mankind has inherited a notion of limitless change. The world has seemed infinitely large, rich and forbearing – even after mankind learned that the earth was finite. Nor are there any global mechanisms for sanctioning local actors or alerting them if they take a wrong turn.

There appear to be two ways to make limits outside the actor's field of vision visible. Expansion can continue until it stops of itself. It's a risky method: the outcome could be catastrophic. Or we could devise theories to help us calculate the tolerance limits of the landscape mantle and devise ways of keeping the succession of local actions within those limits. This is the option which appears to be gaining ground. 'Mankind must determine and respective the environmental limits to growth', writes the oceanographer John Woods in the summer of 1989 in an article on the oceans' renewable resources.

The Cartesian legacy

In the piece cited above, SCOPE refers to the responsibility of the scientific community to study environmental problems. Their results can be used to inform strategic planning. One might be inclined to see the hand of irony at work here, not least because science is behind many of the technological innovations that are responsible for our current precarious state. Wouldn't it be risky then to prescribe the same medicine for the patient's ills?

Happily, science is neither monolithic nor unchangeable. But it is high time to complement the long prevailing analytical perspective with another, a synthesizing one, as a counterweight. Of course, doing it, that is, getting something more practical out of it than the rhetoric to which many 'holistic' advocates are devoted, is easier said than done.

The most successful sciences – physics, chemistry and molecular biology – proceed according to what is sometimes known as the Newtonian programme, whose origins lie in the rule formulated by Descartes: 'divide every problem into as many pieces as possible and necessary to solve it in the best way'. The implied assumption being that the whole is equal to the sum of its parts.

The rules have been followed insofar as we identify and study smaller and smaller pieces of the material world, down to the subatomic level, where it is impossible to comprehend what exists and is going on with the terminology of the macro world. The organizational result is a veritable confusing plethora of specialist sciences, whose representatives find it extremely difficult to combine their results to give us an understanding not only of individual, simple systems but also of the complex systems constituting the framework of our existence.

Knowledge which simply exists but has no technological application becomes a piece of cultural heritage of little practical importance. In that sense, there is no significant difference between contemporary cosmologists' Big Bang and medieval theologians' Day of Reckoning. Both belong in their way to grand systems of thought.

Not so with the type of knowledge used in technology, which is what happens to an increasing extent. In this respect, the micro orientation of the leading sciences is particularly troubling. What it means is that we have a good understanding of technological contrivances in terms of their internal structure and immediate function, especially under laboratory-like conditions. (Factories and certain other closed entities can be seen as big laboratories.) But the direct and even more the indirect impact on society and nature is generally not understood in terms enabling us to undertake an impact assessment. The Newtonian programme has undermined interest in the landscape mantle as the unavoidable intersection where material and cultural phenomena collide. The ecologist I quoted above, Eugene P. Odum (1989), touches on this question in pointing to the problem of 'the increasing specialization and the preoccupation with laboratory study. The open systems of real-world environments cannot be enclosed in glass tubes or laboratory walls'. The debate on genetic technology can be seen as a classic example of the disturbed balance between micro and macro knowledge, where the former conceptions are highly efficient and the latter vague or non-existent. Regarding the risk we are running in this situation, an observer comments: 'Unlike the chain reaction in a nuclear power plant – when the elements are contained and controlled – the release of genetically manipulated organisms could launch a chain reaction which we can neither understand nor control. This chain reaction will not be in the laboratory but in our world' (Fowler *et al.* 1988).

The legacy of Alexander von Humboldt

It isn't difficult to find a prominent figure after whom we can call the programme we need to put in place alongside the Newtonian all the way from school to cutting-edge research. Alexander von Humboldt laid the foundations of modern geography and therefore specialized geosciences as climatology and biogeography. He created, for instance, order among spatially dispersed data using the concept of isotherm. He devoted himself at the same time to population and supplies issues. Above all he emphasized the duty of science to pursue a synthetic approach.

Many of his achievements have of course been rendered redundant by the passing of time. Synthesis is no longer possible using his brand of broad knowledge. It has to be handled by some form of abstract unifying conceptualizing. This is no more remarkable than that time has also rendered much of Newton redundant.

A Humboldtian programme should of course not focus on the history of ideas but on developing a perspective attempting to contain all the deep interdependencies within the earth's landscape mantle within its focus. This is no easy matter given our present conceptual apparatus. Theories are about individual systems. But there are no simple environments apart from those within the walls of the laboratory or factory. In line with the prevailing winds, the eco- and geosciences and social research too have gone a long way down the path of compartmentalization and specialization. One of the consequences is the continued absence of congruence or correspondence between the temporal or spatial scales applied in the natural sciences. On the cultural front we can also note the absence until very recently of systematic research into the role of technology in social and environmental change. Humanists and social scientists have no generally applicable worldview at all with the capacity to facilitate a meeting with the natural sciences.

At a deeper level, model building, in both natural and humanistic sciences, suffers because verbal designations and mathematical symbols deprive things of their materiality. The struggle for elbow room within the landscape mantle therefore resists description and investigation as a process. But it has to be done by some means, in order to enable an assessment of how much mankind can allow itself to appropriate without hazarding its very existence.

As we gain a better understanding of mankind's elbow room, we face the task of keeping our actions within the critical limits. From what has already been said on the primacy of local over global action, it follows that the only way to repair damage and avoid hazards in the future is by changing our approach to local action. We can learn more about the global, we can talk about global issues and we can (possibly) agree on global strategies for change. But it will be the many local actions dispersed across the landscape mantle that produce over time the desired global changes.

Co-ordinating the small decisions

A sustainable development requires a co-ordination of human activities in the landscape mantle of an unprecedented form and reach. A start can be made right away by concluding international treaties and, in their own backyards, national legislation on environmentally hazardous substances and protection of the environment. The scientific communities working on these questions appear, however, to be far too complaisant in believing these verbal symbolic actions actually lead to results on the ground. Words on paper are one thing, practical action in the physical world another.

One primary condition is this: the various localities must be capable of acting in conformity with generally accepted norms. A second condition is the existence of a sufficiently robust community of values and social discipline. These matters vary widely from country to country. They are present to an unusual degree in Sweden. Having said that, the failure of, for instance, the forestry sector (Eckerberg 1987) and agricultural sector (Fries 1988) to adhere to Swedish environmental legislation is an empirically established fact. So what can be expected of developing countries with weak governments and populations on the brink of starvation?

There is a great deal to learn, no doubt, from the regional policy many countries pursued after the war. Regional policy has tended in practice to be enacted politically and administratively as any other sector policy. Consultations with sector representatives have occurred and in some cases resulted in co-ordinated action. But in the main, the function of regional policy has been to repair the damage after technological progress and national economic policy had done their thing. Experience tells us that sectoral regional politics matters very little for the results of regional development. The combined influence of the other policies and general economic output on the geographical distribution of jobs and housing is much stronger.

It seems today as if sectoral environmental policy and environmental management are subordinated in a similar way in relation to all the other forces which together determine human expansion in the landscape mantle. Changing the deep trends to the extent necessary if we want to respect the constraints indicated by the natural sciences, will need much more fundamental behavioural change than what is attainable through a sectoral approach. To satisfactorily co-ordinate the local with the global in a way which can do justice to the name sustainable development, one of the main undertakings will be to think very carefully over the type of institutional arrangement we will need for the job.

According to many observers, unless the general public endorse the new assessments of the situation, the politicians won't budge. While this is clearly correct, it won't be enough on its own. We must be able to convert the analyses into local action in harmony with the wider state of affairs which we want to change. The 'tyranny of the small decisions' is in this respect a fundamental problem. Each separate actor sees and perceives in a direct way only the local situation in which she finds herself. Everything beyond the local, be it regional, national or global, remains on a theoretical level. The wider view depends on knowledge assembled and organized by others. Similarly, the bodies responsible for rules and norms also rely on theoretical knowledge.

So the type of knowledge researchers convey to policy makers is crucially important. And it is easy to identify what needs to be done first of all. We should seek to achieve a better balance than we have managed so far between the diverse forms of knowledge based on fragmentation, and forms able to facilitate a synthesis. It has become particularly urgent in recent years to keep our feet on the ground and attempt to understand relations between the global and local both in nature and in culture. But that can't be accomplished without new conceptualization, which should not be confined by the conventional compartmentalization and specialization in the natural and humanistic sciences. We need to discipline our thinking, but not necessarily with a form of discipline which ploughs old furrows deeper than they already are.

Note

1 The term originates in Soviet geography.

Bibliography

Eckerberg, K. (1987), *Environmental Protection in Swedish Forestry. A Study of the Implementation Process*. Umeå: University of Umeå.

Eddington, A.S. (1935), *Space, Time and Gravitation. An Outline of the General Relativity Theory*. Cambridge: Cambridge University Press.

Eldblom, L. (1968), *Structure foncière. Organisation et structure sociale*. Lund: Uniskol.

Fowler, C., Lachkovies, E., Mooney, P. and Shand, H. (1988), *The Laws of Life. Another Development and the New Biotechnologies*. Development Dialogue 1–2. Uppsala: Dag Hammarskjöld Foundation.

Fries, M. (1988), Naturvårdshensyn i odlingslandskapet. Seminarieuppsats vid Institutionen för kulturgeografi och ekonomisk geografi, Lund: Institutionen för kulturgeografi och ekonomisk geografi.

Kahn A.E. (1966), The tyranny of small decisions: market failures, imperfections, and the limits of economics. *Kyklos*, vol. XIX, pp. 23–47.

Odum, E.P. (1989), Input Management of Production Systems. *Science*, vol. 243, pp. 171–82.

Ratzel, R. (1899), *Anthropogeographie*. Stuttgart.

SCOPE, Scientific Committee on Problems of the Environment 1989–1991. International Council of Scientific Unions, Paris.

Smil, V. (1988), [Book Review of] 'Sustainable Development of the Biosphere'. Ed. by W. C. Clark and R. E. Munn. Cambridge University Press 1986. *Economic Geography*, vol. 64:3, pp. 280–2.

Woods, J. (1989), *Limits to Exploitation of the Renewable Resources of the Ocean*. Paper for Symposium II 'Renewable Resources of our Planet'. Academia Europaea meeting, London, 27 June.

10

ON MAKING SENSE OF THE HUMAN ENVIRONMENT

The problem

Torsten Hägerstrand

(PUBLISHED IN 1977 BY THE ROYAL SWEDISH ACADEMY OF LETTERS AND IN 1978 BY LIBER. SEE BIBLIOGRAPHY AT THE END OF THE BOOK)

In its original cast this paper formed the keynote address at a conference on 'People in the Techno-Society – Approaching Human-Ecology in the Light of a Humanistic Social Science' organized by the Academy of Letters, 25–27 January 1977, at the Old Parliament Building, Stockholm.

Readers may feel they have reason to ask why a keynote speech from a conference on people, technology and society is being printed in a journal devoted to regional politics. The ideas expressed in the speech are, however, of a completely different calibre than questions usually associated with the geographical distribution of industry and housing.

But it can't be denied that anyone who has studied regional development and government for a time and observed reality's resistance to changes imposed by our political masters, might start asking themselves whether the problem was wrongly phrased and the research too shallow. Can we really do more than we already have with the toolbox of concepts and procedures with which we work, researchers as well as politicians and officials? Do we really need to probe the deeper levels of our worldview in an attempt to understand industrialized society's problems, human, environmental as well as regional?

These very questions were raised not more than two years ago at a series of seminars at the Institute for Human Geography, Lund, entitled 'Knowledge and Experience', chaired by professor Anne Buttimer of Clark University, USA. An observer with a different observational background and approach to people and things, was exactly what the doctor ordered to cast in a new light on the provincial and habitual. In a world in which we have fallen prey to specialization's narrow field of vision, it is far from easy to rise high enough above the fray to discern untried routes. While we do indeed hold partial truths in our hands, we'll have to wait for the great architect himself to show us the bigger picture in a coherent fashion.

In the lecture I remarked on certain themes which seemed conducive to my argument. One is that the scientific approach to categorizing and specialization is so entrenched in our consciousness that we have allowed technology and social organization be governed far too mechanically, without letting human experience and reflection provide a sufficiently weighty counterforce. In a regional policy context, it means among other things that the regional problems hinge just as much on our way of categorizing things as on malicious features of 'reality'.

Another theme closer in its way to the preoccupations of regional policy studies, i.e. accessibility in the physical sense – accessibility to jobs and services, in different parts of the country, that people either enjoy or should enjoy. It turns out that the idea of something being within 'reach' has several other connotations which we so far have lacked the wits to give it. Physical access, which we in the social sciences have concentrated on, is often a precondition for actualizing the other dimensions. But the concept 'within reach' could also draw several other things together. It would be very useful indeed for researchers and policy makers seeking to transcend sectoral boundaries to develop the concept of 'reach' further.

Epitaph

There is an inscription on a gravestone in Sweden which the deceased himself wanted us to ponder as we stood beside his grave looking across the lake to tree-clad hills. It goes:

> Here lie the remains of an old man. He gave thanks for the incomprehensible gift of having been blessed to live a life on earth as a human being.

Now it is the case – in our country at least, that never before in our history have so many of those born to the world been blessed to live a full human life and reach a ripe old age. We can't know how many of these people enjoyed life as a miracle and adventure. Quite a few, possibly, since so much of great value has been accomplished. But what's the situation today? Sociologists tell us that for many individuals in the current cohorts of teenagers and their parents, life has no content or point. To them, the rest of their lives aren't a wonder to behold, but a tedious burden.

Sweden is an interesting country, maybe the most interesting in the world. Our material standard of living is incomparably high. At the same time, society is permeated with a fundamental sense of solidarity with the vulnerable, a solidarity usually not depicted as part of the market economy. The economic safety net is nonetheless closely woven around everyone. Life is also, moreover, protected by a myriad of benevolent and considerate rules and norms. Very little is left to chance. Sweden shows, say observers, that economic development can go together with solidarity between social groups. This is why Sweden is often considered a country worth emulating.

The notion of Sweden as a paradigm has not been without its critics. Observers whose main interest is in social practices have asked themselves why it is that solidarity can't be translated into action without so much red tape, destructive of all spontaneity. Those with an interest in what goes on inside people's heads wonder what it could be a symptom of, our ever-present and omniscient tendency to remain silent and serious, as if we were on our way to a funeral. Did *joie de vivre* disappear as a necessary adjunct of poverty's eradication? Or could it be that all personal characteristics are erased through conformance with anonymous social roles? Or have labour unions and government officials, despite having our best interests at heart, made us grumpy and niggardly to the last ounce of impartiality? Or did the atmosphere turn sour when God disappeared? Why then – when we're better off than everyone else – is it that so many find living a life as a human anything but a miracle?

The inner fields

All these questions are remarkably good for stirring up controversy. The nearer we get to our inner fields, the more our own experience tells, the more distinctive our perspectives and opinions. But quite a few of us, though, could nevertheless agree on one thing, that in the midst of all this affluence, a widespread sense of emotional insecurity is crying out for attention. And that the life worlds of many are characterized by listlessness and emptiness. Circumstances have ordained that we in Sweden shall be watched over by others to see how we succeed in migrating from material well-being to cultural well-being in the widest sense of the term. The basic hypothesis then is that social change, in a relatively short time, has destroyed many vital aspects of life, without them having had time so far to re-emerge in a different guise.

Industrialization and urbanization have eroded many of the old, self-evident ties with nature. Most people used to live *in* nature and *off* it. Today, we live *off* it but can't quite see how. My childhood glades were full of flowers every spring, before becoming hay and pasture. The physical sensation and pleasure merged into one. We knew where our morning milk came from. The food I consume today can't be traced further than back to the supermarket. It's impossible to follow the implications of our consumption habits as they ripple backwards along the chain.

Migration and mobility – and sitting on the couch watching TV – have broken many of the tight social bonds between settlements and generations. The small nuclear family turns in on itself. It is no longer attuned to the consequences for others, especially the lonely, of its own isolation.

The collective and individual historical perspective on the past appears to have dissipated into random fragments of more or less the same size, plus a few others. Children today don't understand Elsa Beskow's stories. What's a washing board, and what does it mean to go to the mill to grind? A lot of people, for their own sakes if nothing more, are trying to repair the damage by investigating family history, a popular movement threatening to leave church records in shreds.

The secularization of the physical sciences removed belief in an inner world order, one which cared about people and, at the deepest level, looked after the afflicted.

In all these connections, before they fell apart, knowledge and sentiment, usefulness and emotional balance went hand in hand.

Desolate worldview

The thing we have to contemplate today is a desolate, frightening worldview, without compassion. 'The individual is nothing but a random arrangement of different particles' according to a prominent ecological writer. Our outer earthly world, in an explosion of new connections, is bound up in a global context. The economic dependencies are impenetrable backwards and forwards. The most conspicuous thing is the flood of unsorted news of revolutions, crisis summits and assassinations. The information disappears as fast as it appears. What does remain is mostly a sense of apprehension.

What can be done to establish the expertise and mentality needed to meet the future?

It would obviously not be prudent to attempt the development of a mentality in a culture with the help of science alone, as if we were assembling a technical programme. The visionary ideas – which are possibly what we need above all – cannot be analysed into existence. They have to come from somewhere else. But we can and should study factors and circumstances likely to slow or accelerate the process. The self-knowledge achieved through systematic study is a first step in reviewing seemingly immutable, taken-for-granted ideals and procedures, which on closer inspection have probably slowed our progress towards cultural fecundity and happiness. The conditions of life are accessible to knowledge and change.

'Human ecology' was coined precisely to underline the urgency of learning as much as possible about human existence in preferably all its multifaceted totality. Ecology – the study of the home, house, of existing in the world – has quickly become fashionable. In fact it's being overused. But it's very difficult to find a replacement. Biologists have to some extent taken the concept on commission so to speak.

'The science whose study is interaction in nature is called ecology' (Perarvid Skoog), it has been said. At least in Sweden, ecology in this sense has become the most popular religious outlook of the time. Religion deals with connections and values. So does ecology, at heart. The conservation movement is, perhaps, the great awakening of the age.

The place of humanity in the thinking and behaviour of bio-ecologists is, however, still unclear. The work cited immediately above on bio-ecology – which by the way is very well written and worth reading – begins:

> Mankind has conducted itself as if the world were infinitely large and as if it believed that nature possessed an infinite capacity to heal itself. We can only conclude that we must work harder to understand how nature is constituted, how it works. It is important not least to ensure that what we already know in this area is brought home to one and all – knowledge of the nature on which we rely must incorporated in our curricula!
>
> *(Skoog 1976)*

I would not call this conclusion wrong. Indeed, we all need to take note of it. All the same, it is incomplete and could, by virtue of its incompleteness, be dangerous. There is a great deal of expertise on the matter, even if it is not always used sensibly. But the present, critical deficiencies are at a different level. The point now is to invest more in understanding ourselves as humans and the society we have allowed to develop around us. As long as we lack deeper insight into ourselves and our society, all the knowledge of nature in the world won't help us resolve the dilemma technological progress has put us in.

Bio-ecologists are outraged by the way in which modern technology deals with nature. In their chosen role as nature's envoys, they appear sometimes to want to push humanity out of the way, reduce it to the status of an animal just like any other. We others find it difficult to accept such a view. One could instead insist that humanity is nature's most complex and brilliantly formed phenomenon. It would be hard on the Creator, if humans failed to learn to take care of themselves and the natural world from which they emerged. He would have to write off Earth as an experimental field for evolution's most advanced trial, the creation of the most remarkable of all natural phenomena, the human mind.

We must hope that an ecology with a broader approach will appear than what we have seen so far. Knowledge of the ecological relations won't find its rightful place until concepts have been worked out that accommodate the co-existence of humanity, animals, plants and machines. An ecology like this must be the goal. It might sound Utopian to all the representatives of the innumerable specialist sciences, who realize they have to join in and tremble before the enormity of the integration problem. But one forgets that new, radical concepts can often make sense of what used to be spread and incoherent. Only an ecology of this breadth can inform a policy for caring for nature and humanity. According to American political scientist Ferkiss, what we need is an ecological humanism (Ferkiss 1974).

Bridges are needed

Getting there will be hard. This is because, among other reasons, the science of nature and science of human beings and their societies are essentially different, and will probably stay different. We need a third type of science to build the bridge. The difference has been articulated by the philosopher Alfred Schütz:

> The natural world as investigated by scientists has no 'meaning' for the molecules, atoms and electrons intrinsic to it. The social scientist's subject-matter, however, i.e. social reality, has a particular meaning and significance structure for the human beings that live, work and think in it . . . Through a range of common sense constructions, they have chosen and interpreted the world they experience as a reality in their daily life. It is these thought-objects that determine behaviour by giving it motives. The thought-objects constructed by the social scientist to understand social reality must be based on the thought-objects constructed by people with common sense, who live

> their daily lives within their social world. In this sense, then, the social scientist's concept is, so to speak, a second order concept.

Now, not all social scientists will see the problem like Schütz. Many conceptual models have been made for human behaviour which are of the external scientific type. But the current tendency in social science thinking seems to move along the route recommended by Schütz (Schütz 1967).

Now humans, as a physical phenomenon, can be investigated by purely scientific methods. But they won't get us further than halfway in understanding behaviour and actions. The driving forces exist on another plane, where what we are facing is not material processes but meaning and significance of a cultural or personal nature. A physiologist, for example, can examine a runner's physical characteristics and make a report on his likelihood of running a certain distance in a certain time. But he can say absolutely nothing about what compels the person to make the effort to run in the first place. Other questions have to be asked about people to do with culture and personality, the varying intensity of desires to express their identity or weigh possible rewards.

Given these premises, an ecology with a serious approach to humanity could not be constructed on the premises of the physical sciences. The circulation of energy in the ecosystem – which nowadays is playing such a big role in bio-ecologists' analyses – is certainly a basic, common problem. But in its scientific conception, it won't help us explain why people and their organizations open and close the gates of the energy sluices like they do. That issue has to be understood by understanding how different actors interpret and seek to master their worlds. Without that insight, there can be no hope of changing how the future plays out.

Ethics

Another profound difference is ethical in nature. Biologists don't doubt that living populations maintain equilibrium in their relation to each other by the pitiless removal of the weaker individuals through hunger or predation. In the human world, there is hardly anything people from all cultures agree on more than preventing the death of an individual. We have come a long way in that sense in the advanced industrial nations. And a large percentage of global population growth lies precisely in this relationship. Protection even extends to the life of the vulnerable. But there is more. There is a demand for life to have a qualitative meaning for every individual.

The protection of the individual puts great demands on how the human environment is configured. Technology, at least, targeted that particular objective for many years. But it has also been said that human ecology is only partly given by nature. The artificial elements predominate more and more and that, indeed, may be the rub. Machines and bureaucracies – which are also machines in their way – are beginning to evolve into very peculiar beasts, unknown until the present. They could overwhelm us.

So there is reason at least in the present situation to establish human ecology as a distinct field of study. But let us for pity's sake not draw hard and fast boundaries against bio-ecology. We have to be fully prepared for the two perspectives to converge, and the sooner the better.

Dangerous errors

All forms of ecology seek to understand in a general way life, its origin, span and demise. In a constructed world, which largely describes the human world, vital clues are easy to overlook. Nature works in a trial-and-error mode. Humanity has done the same, so far. But given the enormity of current experiments, the risks are immensely dangerous. More than ever, we have to think before we act. But the main thing is to examine what actually happened before the present. Before we start writing prescriptions on how and what to change, we should ask ourselves where exactly we are. The first step then is to learn more about ourselves, individually and collectively.

One way of acquiring self-knowledge is by thinking hard about which basic ideas we take for granted in relation to humanity, society, nature and religion. What is taken as an axiom at the deepest level in a culture, develops such an ability to pervade the intellectual and practical world, that we can almost start excavating wherever we want. Wherever we start, we will ultimately arrive at the same configuration of the basic ideas.

During work on the conference programme, which the Academy of Letters arranged in 1977, questions like these were discussed. I suggested giving the conference a theme about people in the techno-society. Why did precisely this phrase come to mind so spontaneously? It could just as easily – or even preferably – have been technology in human society. There can't be a techno-society without people – unless robots overtake the planet.

Nor can there be human societies without technology. But the sequence of words says something about how we let our thoughts arrange themselves. It says something about what we believe to be the independent and dependent variable. The phrase 'people in the techno-society' indicates an idea of the irresistibility of technological growth, akin perhaps to the biological revolution, while humanity and other forms of life have no alternative but to get in step, as always. The phrase 'technology in human society' indicates, on the other hand, a belief – at least it can't be anything more for the time being – that technology can be the servant of humanity and life. How it should progress is a carefully considered question. Many of our habitual ideas will have to change before that description is likely to be true.

The main problem is: *what do we do in terms of causal analysis and course of action that will allow us to overthrow the phrase 'people in the techno-society' and arrive bit by bit at a situation in which a group like us can meet to discuss 'technology in human society'?*

If we place the concept of 'human society' first, we need then to understand ourselves and the factors enabling our interaction with each other and nature with concepts of such powerful validity they can help us say yes or no to the technological components and systems which want to entangle us by their promises and their ties. By technological component and system we understand material technology and organizational constructions of a machine-like nature, like, for instance, bureaucracies.

Keyword: reach

We have a responsibility to establish a holistic view of human ecology as a field of study where mankind is centre-stage. This is not an easy responsibility to perform, given the multiple perspectives and terminologies. I have been looking for a keyword, something to underpin the discussion and which we can return to time after time. It needs to help us keep the individual in view as seen from her place within the network of inner and outer connections. But it shouldn't be so naive that it stops our manoeuvrability in time and space, from the intimate to the global.

What I can offer is a single, short word, borrowed from the phenomenologists' and geographers' conceptual toolbox (Buttimer 1976). I'm thinking of what in English is called *reach*, and in Swedish *räckhåll*. A comment is necessary to clarify its significance.

That something lies within a person's reach means she can grasp it physically. But that's not all: economic, intellectual and emotional dimensions are implied as well. Put in pictorial language, every person during their lifetime is enveloped by a *reaching space*, stretching to either side to various degrees, and backwards to the past and ahead to the desired objective. What the space contains depends on the person's abilities and biography, but also on what the environment offers in the form of suppliers to take from and receivers to give to. More precisely, we could speak of *attainable reach*, *manipulable reach* and *restorable reach*.

Conventional welfare policy has largely involved putting more and more material goods and certain services within the reach of more and more people. A foreign observer has compared the results to a box of Lego building bricks. They come in various shapes and sizes and colours, and can be combined in many different ways. On the other hand, though, there has been less interest in people's capacity – possibly even need – to give. Altruism is not a popular enterprise in the eyes of special interest organizations. It slips through the net of the bargaining projects. But even generosity belongs to the idea of reach.

In the rush to increase the availability of certain things, other areas have sustained a loss. I have already mentioned the harm done to the social networks. Moving to a faraway place – not to mention bigger labour markets – incurs costs for the people who move and for those who remain. Similarly, social isolation comes easily to the family whose father works in one place, mother in another, and the child spends most of its day in a third. Even within the family tensions will erupt when workplaces require one or both breadwinners to work shifts.

There are other cost items. Individuals themselves must be capable of absorbing the knowledge and qualifications relevant to their needs and demands of the environment. It is difficult to gain new qualifications at the pace of technological change. Again we see a ruptured connection. Many workplaces and other providers in the environment will then be out of reach for the individual.

The knowledge question

Accomplishing something is one thing. To resuscitate something is something else entirely. There is a relatively widespread assumption – at least, it is something worth investigating – that technological progress has made it harder by and large for the individual to control what is within her reach. Even arrangements of a purely local character are based on highly formalized decisions or constrained by norms determined by disparate authorities. The division of work into small pieces exacerbates the job of identifying and getting in touch with the person who is actually in charge. But again, the biggest obstacle is possibly the knowledge question. Who would dare to have an opinion on something they don't really understand? More and more is within the manipulable reach of the expert, but outside the reach of the ordinary citizen. Is it healthy to live in a world whose construction principles are incomprehensible?

The rivers of information are not really manageable either. I am the target of numerous information merchants, but who do I turn to if I happen to have a question? The TV overwhelms me with information. Information I can't really translate into knowledge. I would have had to go through it all again and ask questions. Real communication requires the communicating parties to be within manipulable reach of each other.

I also think we can say – at least it is another thing worth investigating – that a solid grasp of the past is important for everyone's bearings in the present. To a certain degree, the past should be reproducible both in memory and in concrete, external reality. We can't communicate in the present unless there is a common cultural heritage from which to draw analogies and allusions.

As far as the private past is concerned, the most important thing is possibly that the eldest can recreate it. They have no future reach, anyway. But memory fails with increasing years. Memories need the help of immediate sensory impressions to work. They need the help of the physical objects involved in the situations. I wonder whether we quite understand that when old people make a fuss about being removed to a clean room in a retirement home, it's not just from an obstinate refusal to change old habits. He's being separated from experiences of a past. The means to recreate the past slip out of reach.

Subtly

The things and access to them that I have discussed so far are quite straightforward. There are phenomena of a more subtle disposition, however, but which are also inferred by the notion of reach. I'm not at all sure how I should go about describing them. They are overlaid with taboos. But somewhere there's a need to access symbols, or at least peace and quiet of the type about which 'the random arrangement of particles' – which constitute my person after all – is anything but indifferent.

It is often tacitly supposed that things could only get better and better as increasing quantities of ready-made goods are placed within the reach of everyone. The bigger the Lego brick box, the better. This is where the idea of growth and standardization comes from. That is a truth with modification. First, there is the question of tolerance. From the individual's perspective, the world and its numerical expansion is an increasingly incomprehensible universe. The number of material aids and commodities is growing, the amount of information is growing, the amount of scientific and technological knowledge is growing and bifurcating, the number of rules and provisions enveloping life is growing. For the individual, things move beyond his reach by virtue of quantity and subdivision. There is certainly an upper limit to how much we can master in a satisfactory way, not to mention individual variations. We need help to keep it away from us, everything that is intrusive, exhausting and unfathomable.

Then there are the contextual reasons. People need to make sense of their world. It's not just a question of quantity. If it were, a hundred-strong orchestra would play better music than a string quartet. Perhaps the child within shouldn't have such a big box of ready-made Lego bricks, but a toolbox and a box of colours and something to mould into shapes.

Isn't it plausible that human society will have to release the creative capacities of everyone sooner or later, and not just among the elites who create the big systems? If so, the impact on the scale of human activity would be enormous. A combination of the small-scale and diverse is a cultural model worth pondering.

Technology has invaded

What I want to do with this reach-related argument is to show a way of exploring how we managed to allow technology, the material as much as the organizational, to invade human society. Things seem to have developed in such a way that whatever lies within reach is too one-sided and whatever lies within manipulable reach and restorable reach is insufficient. Do we need more evidence of the irrational invasion of human society by technology than calls to keep day care centres open at night as well? Children sometimes have bad dreams, and they have a right to a familiar and ordinary adult bed to creep into. Why have we reached the point in society's evolution that we can no longer offer our children a safe night's sleep?

I have attempted to suggest one possible way of keeping a handle on the discussion of symptoms and hopes, even perhaps an embryonic way to bring together knowledge from different sources. This is a chance to move forward and get down to causal analysis.

It is hard to dismiss the idea that the current structure of scientific knowledge is in itself to blame for the situation we have let ourselves in for. If that is the case, we should start examining our own outlooks and conceptual structures before advising others.

It wasn't necessary to let things get this far, unless there was some basic deficiency in the conceptual categories we apply when we observe, analyse, produce, administrate and politicize. It goes to show how much hangs on the axiomatic categories.

Within the space of a couple of centuries, thinking has become completely dominated by the principles of dividing responsibilities and specialization. The principles have obviously been immensely useful by multiplying industrial output, increasing scientific knowledge and widening personal freedom. During the process of specialization we have concentrated on devising smaller and smaller and more and more bits of the natural environment, technology and society. Experimental technology and statistics helped us establish regulated connections between small groups of variables. The welfare state adopted the same construction principles. As soon as a hitch appeared somewhere, we were on to the problem and put in countermeasures. Bureaucracies have often been created just to deal with this factor. Thus, progress has expanded and spread the technological machinery and social machinery.

Hitches

The hitches, though, are spreading. The friction is most apparent in the health sector, where an over-exercised component philosophy in some way or other is contrary to its nature. It is as if some sort of abstract space had been filled to the brim with components and component systems, each individual component entirely sensible in its way, but together with all the others, now entangled in all kinds of unforeseen mutual discord. There may even be reason to fear that the social system has become overdetermined, that we will find it extremely difficult to adapt to new lifestyles if required by a crisis in the international situation or simply because that is what we want.

Progress in the field of specialization and division of labour has caused us to overlook things of a certain magnitude which can't be cut up into small pieces and handled haphazardly. On the one hand we have the indivisible individual, who neither can nor will cut themselves into however dopey and incoherent a pattern of whatever roles and responsibilities. On the other, we have another totality, the biophysical population/environment system, which similarly cannot cope indefinitely with technological and organizational stress.

But if we proceed to the construction of the sciences and conceptual system, we find two things. The individual person and factors in her total and indivisible life world have vanished through the holes in the net with which we try to catch reality. Novelists have been left to grapple with this perspective more or less on their own. But precisely the opposite has affected the overarching social structures, which inform the general direction of history. It is to the Marxist researcher's credit that we are aware of the structural cavity. The human dimension has, however, also been neglected by them.

It is clear to me that human ecology has two levels of integration. The one consists of the individual, indivisible and with a right to make sense and find meaning in every part of their internal and external world. The second consists of the total, interactive assembly of people, organisms, things and signals, whose forms and processes will determine whether the individual life is a miracle or a misery. By hook or by crook, we have to create thought structures for these two levels and put these concepts to right. A more accommodating attitude to interdisciplinary cooperation is a first step. But not enough on its own. We have to make sure that the field is free so that we can approach reality in completely novel ways.

The attempt is worth making and making precisely here in Sweden. A foreign observer (Buttimer 1976) put it like this:

> It seems to me as if the challenge Sweden is facing, different more in degree than substance to the one facing other affluent countries in the West, is to foster a productive, caring and responsible society, in an environment without material poverty. If you opt for this demanding course, your contribution to European history will be enormous: to revive is more admirable than to innovate.

Bibliography

Buttimer, Anne, Knowledge and Experience. Unpublished contribution to the seminar series. Lund, 1976.

Buttimer, Anne, *Home, Reach, and the Sense of Place*. Uppsala, 1978.

Ferkiss, Victor, *The Future of Technological Civilization*. New York, 1974.

Schütz, Alfred, *Concept and Theory Formation in the Social Sciences. Readings in Existential Phenomenology* (edited by N. Lawrence and D. O'Connor). London etc., 1967.

Skoog, Perarvid, *Ekologi*. Svenska Naturskyddsfdreningen. Uddevalla, 1976.

11

HUMAN RIGHTS AND ECOLOGY AS PREMISES FOR TAKING PRACTICAL STANDPOINTS

Jon Wetlesen

(THIS TEXT IS AN ABBREVIATED TRANSLATION OF WETLESEN'S NORWEGIAN-LANGUAGE ARTICLE 'MENNESKERETTIGHETEROGØKOLOGISOMPREMISSER FOR PRAKTISKESTANDPUNKT', ORIGINALLY PUBLISHED IN 1975 IN A. VINJE AND P. HOFSETH (EDS): *ØKOLOGI – ØKOFILOSOFI*. OSLO: GYLDENDAL)

Introduction

This chapter is an attempt to clarify a possible set of normative premises and combine them with descriptive premises derived from ecological and, in part, the social sciences. These premises are then used to reach some relatively general positions on practical, ecological problems, mainly problems related to policy on natural resources and population growth.

Since the original reason for writing this chapter was for a lecture for students studying examen philosophicum [university entrance exam], I decided to use a selection of normative premises less likely to create controversy politically speaking, i.e. the United Nations norms of human rights. These norms are widely accepted in political circles today, at least at the verbal level, and could be useful as a starting point for rational eco-policy making.

Some might have preferred me to draw my normative premises from among the worldviews of political parties, philosophies or theologies. While that might been more interesting for the adherents of those views, their opponents might have been less impressed. The advantage of human rights norms is that they can be justified from very different philosophical positions. Antagonistic parties can agree on them to a relatively large extent, even if they dispute the deeper philosophical or theological underpinnings.

The main part of the chapter argues the case for certain relatively general conclusions relating to the politics of ecology, while the final section explores ways of enacting an ecological policy of this nature. As a source of norms, I use the 1948 UN Declaration of Human Rights and its subsequent specification in two international conventions of 1966 on economic, social and cultural rights and on civil and political rights. These rights are summarized in norm N1 (p. 148) and define a justice and welfare paradigm for the world as a whole.

In the main part of the chapter I shall be using the economic, social and cultural

In the main part of the chapter I shall be using the economic, social and cultural rights as normative premises. They are set out in section A and combined in section B with descriptive premises of an ecological nature. The conclusions are given in the shape of nine general norms.

The last sections outline the practical execution of these standpoints in everyday life. I shall not be addressing the strategic or tactical questions in detail, but I shall mention certain supplementary premises of relevance. Normative premises are drawn from the UN's civil and political human rights, and descriptive premises from the social sciences. I shall comment briefly on parliamentary and non-parliamentary incentives and mechanisms.

Arguments on behalf of certain eco-political positions

A. Normative premises

Conflicts of interest and ethics

When attempting to make sense of methods of making and justifying ethical choices it helps to approach them as methods of resolving conflicts of interest. If the interests of two or more parties are incompatible – say they all want the same limited natural resources – it will usually be in their interest to explore non-violent means first, before resorting to force and violence. This is particularly pertinent when resolving eco-conflicts, not least because resources in most areas are getting scarcer and scarcer.

One area of particular interest today is precisely eco-politics. When devising international or national legal norms on resource use, pollution, population policy and the like, there is a strong inclination to take one's cue from the moral norms set out in the UN Declaration on Human Rights. There are several reasons why these norms are good at the job. One is their universal character. They concern and apply to all people, today and in the future. Since many of the most important ecological conflicts affect large numbers of people, both in affluent and poorer parts of the world, and both in this and possible future generations, there is a need for norms based on a universal premise. I shall briefly summarize these norms in keyword form under the following main norm:

> NI Practical Main Norm
>
> In all forms of decision making, whether individual or collective, private or public, preference should be given to such options as are compatible with requisite standards of welfare and justice as defined by the UN's human rights' norms.

The UN's human rights norms

There is widespread support in public opinion for the human rights norms in the world as a whole. The norms have been institutionalized more or less comprehensively by many different organizations and have won the approval of the participants in these organizations and the wider public. The organizations are both public and private, of all sizes, and international, national and local in scope.

One must naturally assume that the apparently broad agreement on the human rights norms may largely be for show. People can declare themselves to be in agreement with the norms verbally, but find fault with them when it comes to practising what they preach. There are several reasons for this, one of the main ones being the ambiguity and opaqueness of the wording of the norms. Interpretations will vary when they are put into practice. They are also misused rhetorically to disguise questionable political interests and influence.

While these factors do reduce the significance of the human rights norms, they do not invalidate them completely. In terms of historical development, we could say they are maturing. Indeed, they have developed considerably over the past couple of centuries, and their content is becoming increasingly concrete and precise in response to practical trial and error.

The economic, social and cultural rights

The economic, social and cultural rights are relevant to the type of welfare to which all people are entitled, ethically speaking. They prescribe certain minimum standards to realize the good life. In that all people enjoy certain minimum standards to the same degree, they speak at the same time to the just distribution of welfare. They prescribe in other words certain minimum standards for realizing the good society, not to mention the good world. All people enjoy the same minimum rights to take part in working life and consume primary goods such as food, clothing, housing, social services, health services, education and culture. Since these norms are anchored in essential or nearly essential human needs, they prescribe the welfare dimensions with which most people would concur. True, they do not exhaust the list of the goods comprising the good life and good society. But they do prescribe necessary, if not sufficient, conditions for such a life and society. It is worth noting that they do not apply to consumer goods alone, but also the right to take part in economic life; they apply not only to physical welfare, but also cultural.

Before these and similar norms are applied in practice, they need to be fleshed out and enumerated. It is not enough to determine the different welfare dimensions, a precise minimum standard or floor must be assigned to each of them, along, perhaps, with a maximum standard or ceiling. Several of the UN's special organizations are currently devising and concretizing the standards. We can mention for example the International Labour Organization's recommendations and conventions in the area of work and the Food and Agriculture Organization's nutrition standards. These latter standards prescribe the minimum physiological requirements of a person in respect of calories, protein, vitamins, certain minerals, etc. It should be possible to incorporate variations arising from sex, age, employment, climate, and the like into these standards. UNESCO's standards for elementary and higher education could also be mentioned.

Work is ongoing in many other quarters today on various welfare indicators or social indicators as they are also called. By defining a floor for each indicator, it would be possible to calculate the percentage of the population 'under the floor', or in the case of a ceiling, the percentage of the population 'above the ceiling'. These figures could then be taken as indicators of the level of under- or over-development or optimum development in a society.

The civil and political rights

The civil rights are enjoyed by all people in virtue of being human, while the political rights are enjoyed by all in virtue of being a citizen of a given state. I shall desist from quoting the relevant articles, providing instead a brief summary.

The civil rights include among other things the right to life, freedom from torture and inhuman treatment, freedom from slavery and forced labour, legal protection, freedom of movement, equality before courts and tribunals, freedom from retroactive legislation, freedom of conscience and religion, freedom of assembly and of association, including the right to form and join trade unions to protect one's interests, right to marry and form a family, and children's rights as well.

The political rights include among other things the right of any citizen to participate in public life, directly or through freely elected representatives, to vote and be elected at periodic elections based on universal suffrage, in secret, and therefore as an expression of the free will of the voters, and in general to have access to equal conditions to serve one's country in public office, equality before the law, and minority rights.

These rights define, among other things, a framework for democratic decision making. They are in consequence relevant for a democratic eco-policy. I shall return to this in the second part of the chapter.

On equality

We come now to a matter of crucial importance regarding the human rights norms: their scope or applicability. Among other things I refer here to the classes or categories of persons (or entities) to which the norms nominally apply.

> The purpose of the UN, as set forth in its 1945 Charter, is to promote respect for and allegiance to the fundamental rights and freedoms for all without distinction as to race, sex, language, or religion.

Similar specifications are given in the conventions of 1966.

In specifying the scope of the rights norms in this way, it ensures a robustly egalitarian distribution of the minimum rights. The logic in essence goes something like this. Let us take one of the most general norms of fairness, i.e. that prescribing the equal distribution of wealth. The wealth to be redistributed can be positive or negative, such as benefits or burdens, duties or rights, power, privilege, prestige, etc. In the present context, the norm could be articulated as follows:

> N1a When scarce common resources are divided, they shall be divided equally among equal recipients, in other words, no differential treatment is permitted unless indicated by material differences of relevance among the recipients.[1]

I call this norm N1a to indicate its kinship with the aforementioned norm N1. The same applies to N1b–N1d below. Together, they provide points of reference for a possible interpretation of the human rights norms.

N1a is a deep-seated norm, and a fundamental precondition for practical reasons, at least insofar as they concern the distribution of wealth. Identical cases shall be treated identically. But why? That it could be any different seems counterintuitive. If two cases are recognized as being identical, there is no reason to treat them differently. For such a reason to obtain, some material difference of relevance would have to be established. But in that case they would not be identical. If they are treated differently despite being identical, their treatment is devoid of reason, i.e. of grounds or *ratio*. It is therefore irrational. One could imagine, for instance, a distributive situation in which one of the parties influences the allocations such as to benefit himself because it serves his particular interests and because he has the power to do so. The division would not be based on reason, but on power.

In vying for society's limited resources and society's economic surplus, the interests of the strongest tend to come first. That's how things have always been in virtually every society in history. As an element in its marketing strategy, the prevailing party will often seek to legitimize its ascendancy, privileges and prestige by appealing to a doctrine of justice or ideology which they attempt to get the less fortunate parties to accept. A key component in any legitimization exercise is the effort to justify why treating different classes of recipients differently is acceptable. A litany of supposedly relevant factors have been proposed down the centuries, including race, sex, language, religion, colour, political and other convictions, national or social origin, birth, property, status, etc. If these reasons are accepted, the result is a class-divided society mirroring the adopted redistribution formula.

What characterizes the UN Declaration of Human Rights and conventions in this area is their critical approach to all such supposedly relevant differences. They are declared irrelevant. All people, in other words, are accorded the same minimum entitlements under the stated rights. From this it follows that the burden of proof lies with those in favour of differential treatment in the redistribution of limited common resources. The pertinent question therefore is this: is there any way at all of justifying differential treatment?

It might be interesting in this connection to refer to an argument formulated by John Rawls.[2] Put somewhat bluntly, he seems to envisage that differential treatment can be reasonably justified if it benefits all of the implicated parties, including the least favoured party. In other words, differential treatment can be justified in this way with regard to civil and political rights. All of the involved, he maintains, enjoy the same minimum claims. Justifying differential treatment in the area of economic, social and cultural rights is something Rawls sees as a possibility. Whether differential treatment is justified or not in respect of these rights must be explained in each individual case to the least favoured of the implicated parties. This, I believe, is a relatively sound exposition of what Rawls means. I can summarize this interpretation in the following norm:

N1b Everyone shall enjoy the same civil and political rights, and everyone shall enjoy the same economic, social and cultural rights, unless disparity in one area benefits the least favoured party.

If we accept this norm, accepting the follow two would seem intuitively reasonable:

N1c In the distribution of scarce common natural resources, *everyone* should receive whatever is necessary to fulfil their minimum rights as defined in the human rights before *anyone* receives more than the minimum to which they are due.

N1d In the distribution of scarce common natural resources of relevance to the fulfilment of the human rights, no one should receive more than the minimum to which they are due unless such differential distribution can be justified as being in the interest of all, including the least favoured party.

These norms are extremely radical. If accepted and practised in contemporary Norway, what would the consequences be? Could we as Norwegians persuade the least favoured parties of our generation in the international community that our current standard of welfare and our current growth rates are justified? And could we do the same to later generations whose resources we are depleting?

This way of looking at the problem is characteristic of the theory of the welfare society. Historically, norms like these have largely informed the distribution of welfare within the individual nation-states or regions. The UN norms would appear to embrace the entire international community and the endless number of future generations under notion. The norms encompass, in other words, not only the welfare states but a welfare world.[3]

Now this could be countered by claiming that justice in the international community is a question of reciprocity or fair exchange, not distributive justice. Distributive justice, in light of the perspective we have just outlined, requires among other things that natural resources are seen as belonging to humanity's common heritage. But things aren't like that in the real world of politics. In the political world, reality is not like that at all. There, resources belong to individual sovereign nation-states, and then only to the extent of their capacity to have their way.

It is, of course, an important objection. All the same, whether it applies to the preceding argument is open to question. As it stands, the objection says something about how rights to property find expression in political terms, while the preceding argument deals with the ethical question of how they should or must be construed in order to be rationally justifiable.

The rights of future generations

We read then the economic, social and cultural human rights norms from the vantage point of universality. We also read them primarily as egalitarian norms. Everyone has as much right to the benefits prescribed by the norms, wherever and whenever they happen to live. We hold the possibility open for certain distributive inequalities, providing they are justifiable to the least favoured parties today or in the years ahead.

One question that has received much too little attention concerns the size of the class of people, or rather of interested parties with an equal claim on the limited natural resources. We need to know this, at least approximately, before we can determine the division to which each interested party has a rightful claim. It will depend on how many people live per year or per generation and how many years or generations are included in the calculation.

The next question requiring an answer is this: how extensive are the natural resources to be shared? As we shall see, the answer to this question depends on whether we are talking about renewable or deposited resources. It will also depend on the type of technology required to utilize these resources.

Although we can only answer these questions hypothetically, we can use hypothetical estimates as a guide for the distributive questions. For each set of estimates of the size of the recipient class and of the commodity to be distributed, we can discuss what would constitute a fair share for each recipient.

How many generations should we assume? The more we include, the lower the yearly resource limit on extracting the resources in question, and this in turn will trigger a number of other consequences for our production, exchange, distribution and consumption of goods and services, and indeed for the reproduction of humankind.

The question can be phrased either as a descriptive, empirical question or as a normative, ethical question. If it is taken descriptively, it could be reformulated like this: how many generations of people are likely to survive in the years ahead? If taken normatively, we could rephrase it like this: how many generations have an ethical right to survive?

We can get a few pointers for the first question from the earth's history. The earth is about four billion years old. There has been life here for about three billion years. The dinosaurs survived throughout the entire geological middle ages of the earth, a period of 130 million years. They succumbed under the impact of climatic change; were we just half as adaptive as them! Our human ancestors walked the earth about fifteen million years ago. The more or less widespread use of simple tools started about 1.5 million years ago. The present volume of the human brain wasn't reached until about 200,000 years ago. *Homo sapiens* can trace its forbears back about 50,000 years or so, even a bit further. It's about 10,000 years since Scandinavia became inhabitable after the last ice age. That was when the agricultural revolution was sweeping across the Middle East. And it's about 225 years since the beginning of the Industrial Revolution.

If we turn now from the past to the future, how far can we see? How many years the human race can survive as a species depends partly on itself, and partly on other factors. Unless we eradicate ourselves from the face of the earth, directly by nuclear, bacteriological, chemical or other weapons, or indirectly by sapping vital resources, we should have a chance of surviving at least as long as the hunter-gatherer societies, that is, more than a million years into the future, even a lot longer! As a prop for the imagination, we could compare ourselves with the dinosaurs. We'd have to extend our planning horizon to 130 million years. That being the case, given our limited resources, a sense of trepidation would seem called for.

How we actually do survive as a species will depend very largely on how we choose to tackle the ethical question about how many future generations we should take into account in our own exploitation of the earth's resources.

Let us look at the ethical side of the question. How many generations have a right to survive? Can we answer this by giving a finite number of generations? If we keep to the ethical aspect of the question, it doesn't look as if we can. If every generation has the same right to have children and see its children grow up, then the ethical criteria won't allow us to set a limit. To do so there would have to be an ethically relevant difference between the final generation and the ones before it. But does such a difference exist? Highly unlikely! It is difficult to imagine how a limit could be drawn without it being ethically arbitrary.

On the other hand, a limit will doubtless be imposed at some time in the future. But it will be a limit set by nature, not one people can determine on ethical grounds. Ethically, we have a right and duty to keep going as long as possible.

The rights of non-human beings

Human rights apply to human beings as a class. But what about non-human beings like other animals and plants, or more complex systems like populations, ecological societies or ecological systems? Don't they have similar rights?

There are both positive and negative answers to the question. I shall outline two of the possible answers. According to the first, rights can be ascribed indirectly to non-human beings, that is, to the degree necessary for compliance with the human rights. According to the second, rights can be ascribed to non-human beings directly, quite simply by extending the scope of some of the human rights.[4]

Let us imagine someone makes the first argument. Whether it is right or not to preserve certain non-human beings or systems becomes then a question of whether it is useful to preserve them in the sense of facilitating the realization of human rights for all human beings over the longest possible term. But, it could be argued, human societies can only maintain equilibrium over the longer term by maintaining the equilibrium of the ecological communities and ecosystems in the environment. Since ecosystem stability requires multiple populations of plants and animals and a complex structure, it would be right for humanity to preserve this variation and this complexity. But it sets limits on how far human beings can simplify ecosystems for their own productive purposes. In this sense, human rights shield large parts of the ecosystems. Non-human beings and systems enjoy in this way certain rights, even if they aren't absolute. But ascribing them depends on their usefulness in relation to humanity's own, easily acknowledged interests when such interests are based on wide-ranging and long-term impact assessments. This is undeniably an anthropocentric form of argumentation.

The other form of argumentation is not anthropocentric. It could be characterized more readily as primate-centric, mamalia-centric, vertebrae-centric, cosmocentric or the like, depending on the extension of the class of non-human beings or systems included in the area of applicability of the rights norms. This mindset, it would not be wrong to say, is more widespread in Indian and Chinese civilizations than Jewish, Arabic or European. It possibly finds its most extreme expression in the Indian religion of Jainism. The Jainist ethics of non-violence comprehends at least the class of all living creatures, including insects. The discovery of bacteria and vira must have posed certain problems. Buddhism and Taoism venerate trees and other plants, not to mention mountains and lakes. In the West, similar sentiments were shared by early Celtic Christianity, Francis of Assisi, and Albert Schweitzer. Schweitzer interpreted his own norm, 'Reverence for Life', in a way not dissimilar to the Indian doctrine of non-violence.

Both the aforementioned modes of argumentation can be used in defence of standpoints favourable to the preservation of non-human beings and systems. Whether one opts for the one or the other will generally depend on one's cultural context. In Norway, the first tends to be the natural choice. Based on its premises, one would not say that animals have rights which are not determined by people, but that people establish a set of norms through which they take on an obligation for themselves and others to care for animals and plants, including complex natural systems, in certain ways. However, the other form of argumentation is evidently advancing in the West today, as a counterweight to a one-sided utilitarian attitude to nature.

B. Additional descriptive premises and eco-political conclusions

I shall now address the descriptive premises and derivations of the normative eco-political views for which I believe these premises give the best justification.

The problem to be dealt with in the following can be outlined as follows. If we assume that all contemporary and future human beings shall enjoy their basic rights as delimited by the economic, social and cultural rights norms, and that the common natural resources (particularly non-renewable resources) one can use to the end, are limited, what would be the consequences for an ethically justifiable allocation of these resources for the choice of technologies, for the type and magnitude of production, exchange, distribution and consumption of goods and services and for the size of the population we should be planning for?

I shall first introduce certain concepts to be used in the arguments, such as renewable and non-renewable resources, soft and hard technologies, a balanced society and its alternatives. I shall then outline certain aspects of the economic and demographic growth in our history before asking what would be the optimum population and economic level in light of human rights norms. Finally, I shall outline some eco-political principles, especially in relation to resource and population policy.

Renewable and non-renewable resources

When we consider nature as a resource, we are applying a utilitarian perspective. We approach natural resources as a means or repository of means for use in achieving human ends. It is usual to distinguish between renewable and non-renewable resources. The Norwegian eco-political debater Hartvig Sætra termed these two types of resources 'circulating resources' and 'stored resources', respectively.

More specifically, renewable resources can be defined as resources which, within a definite, shorter period, are completely replenished, while non-renewable resources can be defined as resources which, once used, are gone forever.

Examples of renewable resources are earth, water, air, fauna and flora. Renewable resources usually come from organic material in an ecosystem. They meet their energy needs by absorbing solar energy.

Matter and energy cycles in the ecosystems can be put to productive use for humans in two ways, either by reproduction or extraction. Reproductive use means that no more is harvested per annum than the resource can afford without sacrificing equilibrium. If we're talking about specific populations of plants or animals, only the yearly population surplus should be harvested. Optimum harvesting is equal to the annual level of growth of organic matter at which the population can remain at a zero growth level. This would normally not deplete the earth's supplies and would cause only minor or no pollution problems at all. Extracting renewable resources beyond the level of reproduction means that more than the annual increment is taken out. This of course depletes the actual resources. Using reproductive resources can be compared to living on the interest of capital, and extracting these resources beyond the level of reproduction with spending the capital. In the former case, resources are used carefully.

Examples of non-renewable resources are minerals, including ores and fissionable matter, fossil fuels such as peat, coal, oil and natural gas.

Non-renewable resources can only be used as raw materials and sources of energy by extracting them. Stocks are tapped. But the amounts extracted can be more or less prudent or reckless. Given our current annual extraction levels and production growth, the reserves of the most important mineral raw materials will not last very much longer, 300 years perhaps for iron ore, 200 or so for bauxite, fifty at a stretch for copper, and perhaps 150 for nickel, perhaps 2,000 for hard coal and brown coal, and perhaps a century for oil and gas, etc. Figures like these are naturally highly tentative. They are estimated on the basis of assumptions of future consumption rates, mounting uncertainty regarding reserves and growing extraction costs. But even if the figures were significantly higher, the resources will clearly not last for any appreciable length of time compared with the time scale we adopted when discussing the rights of future generations. All in all, annual consumption rates will arguably have to be cut drastically.

Soft and hard technologies

By soft technologies I understand here renewable and low-energy technologies. The former characteristic applies to the materials technologies make use of, the latter to energy consumption.

Renewable technologies seek to recycle materials used in industry to make them last as long as possible. At the same time, they try to avoid environmentally hazardous goods and waste products, i.e. by preventing excessive emission of non-degradable and other substances such as toxic matter. They are said to rest on the principles of endurance and recycling.

Manufacturing based on renewable resources means returning materials to the ecosystems from which they were taken. Sewage, for example, should not be allowed to flow into the sea, and dead leaves should not be burned. Instead, we should use different types of compost. With regard to industrial production based on non-renewable resources, it means making things which last and recycling materials as much as possible in the manufacturing process. Since a certain percentage is always lost in every production cycle, materials would clearly be used up.

Progress in the field of renewable technologies will include among other things reducing wastage rates in the use of non-renewable resources.

Low-energy technologies are based in the main on traditional sources of energy known from pre-industrial societies: muscular energy of humans and animals and the three Ws: wind, water and wood. Wood is used for fuel, for example, wind for sailing ships and windmills, water for water mills and river transport.

I take hard technologies to mean technologies requiring the consumption of large quantities of metal and power. Modern industrial societies rely above all on hard technologies.

Since the Industrial Revolution energy consumption has grown exponentially. In 1971, the world as a whole used 7,260 million tons of coal-equivalents, very nearly 300 million tons more than in 1970. Of this amount, oil and gas constituted 64 per cent and coal 34 per cent, while hydro and nuclear power made up 2 per cent. In other words, 98 per cent of consumed power came from the fossil fuels, or non-renewable resources, while a mere 2 per cent was obtained from renewable resources.[5]

The growth in energy consumption per country is very uneven indeed. For instance, the United States, which is home to 6 per cent of the global population, consumes 30 per cent of the earth's power in the space of a year.

The difference between soft and hard technologies, as they are defined here, correlates with several other characteristics. According to a UNESCO typology, differences can be plotted across thirty-five dimensions, including the following:[6]

The soft technologies are adapted to the ecology. Energy consumption is low, pollution is non-existent or minor, they recycle materials and energy, they operate with a long-term time scale, they stress solid workmanship, are not particularly specialized, they go together with extended families, occur mostly in agrarian societies or rural towns and villages, are integrated with nature, have democratic decision making procedures, technology stays within natural limits, and goes together with local trading, sustains local culture, fits in with regulations imposed to limit the misuse of technology, relies partly on other living creatures apart from human beings, innovation is motivated by human needs, they are in harmony with zero growth, are labour intensive, bring young and old together, are decentralized, have advantages for small-scale production, operations are easily comprehended, work-related accidents occur infrequently and only on a small scale, they can find several different solutions to technological and social problems, they go together with farming, accentuate quality more than quantity, everyone participates in food production, work is satisfying in itself, not just as a means, the economic units are self-sufficient and small, science and technology are part of the culture, everyone practises science and technology, there is little or no difference between work and leisure, the concept of work doesn't exist, the technological ends are accepted as applicable to all people all the time.

The hard technologies are the reverse on every point. They are ecologically hazardous, energy and material intensive, polluting, short-term, highly specialized, require cities and global trade, destroy local cultures and other living creatures and ecosystems, are centralistic, authoritarian, etc.

The steady-state society and its alternatives

I shall now introduce a new term, the steady-state or equilibrium society. In doing so it is important to note the definitional characteristics of the term and accessory terms. I select three sets of characteristics: one concerns production and consumption in society, one is related to the reproduction of society and one involves its constancy. I suggest the following delimitation.

A society can by definition be said to be in equilibrium over a given period if capital and population growth is zero and at a constant level throughout the period, that is, its production rate corresponds to its consumption rate and its birth rate corresponds to its mortality rate, and that these rates remain constant.[7]

We need to include a third characteristic as well; it concerns rate stability. If the rates rise, but balance each other, we have a steady-state consumption and production level, but not a steady-state society in the sense I have defined.

This term is especially adapted to human steady-state societies, but contrived according to the same principles that apply to all types of ecological steady-state societies.[8] There is also a likely causal connection between a balanced human society and equilibrium in the surrounding ecological communities and ecosystems. To maintain equilibrium in the former, equilibrium must be sustained in the latter. This, however, is a proposal or hypothesis on the conditions for realizing a human steady-state society, not a definitional characteristic of such a society. The same can be said about other characteristics of steady-state societies, insofar as they display, for instance, high levels of differentiation and integration, complexity and structure, diversity, etc.[9]

By a society's capital I mean here the results of past work which can be channelled into production at a later date. I'm thinking here mainly of real capital, that is, physical goods. That a society's capital growth is zero and constant over a given period means that as much is produced in the period as is consumed privately and publicly, and also used for offsetting capital depreciation. There is, therefore, neither capital accumulation nor capital depletion in the period. The economy is kept in a state of rude health.

Zero population growth at a constant level throughout the period means that equal numbers of people are born and die during the period, or more accurately, birth and immigration rates offset mortality and emigration rates. Population growth in the period is therefore neither positive nor negative.[10]

In light of this definition of the steady-state society, determining the rates of production and consumption, births and deaths remains to be settled. In general one can say that they higher the rates, the higher the rate of flow through the system. In other words, the higher the consumption, the more the production of goods will have to increase to keep society in balance, entailing higher consumption of resources. Logically, then, the longer we want to maintain a steady-state society at a given consumption level of specific, limited non-renewable resources, the lower the consumption rate of these resources, which in turn will decide optimum population size, assuming that every individual enjoys a fair minimum share of this consumption. It is in this sense people talk about an economy for Spaceship Earth.[11]

Natural and artificial population control

The crucial question in respect of population growth is this: what sort of factors will regulate growth? Can it be done by lowering the birth rate and raising the mortality rate, or perhaps a combination of both? If effective birth control can't do it, can it done by controlling mortality?

We can include under the mortality regulating mechanisms 'the horsemen of the apocalypse' proposed by the Revelation of St John 5:1–8, i.e. famine, pestilence and war. We can also include procedures used at different times and places such as abortion, infanticide, geronticide, the killing of disabled, prisoners of war, etc. The birth controlling mechanisms include the traditional approaches such as abstinence, often presumed to be fostered by celibacy, higher marital age and the like, 'safe periods', *coitus interruptus*, mechanical and chemical methods of prevention of various types, sterilization, etc.

Another classification of population regulating factors draws a line between so-called natural and artificial procedures. This particular form of classification is probably more prevalent in the Catholic Church than elsewhere. Among the artificial procedures we can include abortion, sterilization, and mechanical and chemical prevention techniques. The natural procedures include sexual abstention, celibacy, 'safe periods', famine, disease, etc. Whether the natural birth regulating factors will be sufficient to control population growth is another question. If they are not, we are faced with a choice between letting natural mortality regulating mechanisms take over or relying on artificial birth control.

Eco-political consequences

The above argument would seem to indicate a reduction the global population over the longer term. Aiming for a stationary population of 8–16 billion will probably not be enough. In the long run, the figure will have to be considerably smaller. If the population is not reduced by effective birth control, better death control is more likely to do the job. Weighing these two alternatives in light of the human rights norms under N1, the former alternative is clearly the only ethically justifiable one. It follows, therefore, that eco-politics should reckon with the long-term objective of bringing the population down to an optimum size at which it can be kept in balance for as long as possible and at an economic level which meets the economic, social and cultural rights stipulated under N1.

As a guide for a long-term eco-policy of this nature, the following norm suggests itself:

> N2 Eco-political main norm
> In all decisions involving resource or population questions, the option most likely to promote a steady-state society over the longer term should be chosen, such that society stabilizes at a constant optimum population size and at an economic level that satisfies human rights norms under N1.

In this formulation of the eco-political main norm, the stress is on resource and population questions first and foremost, rather than pollution issues. The former are arguably more essential. If we can control them, we will also control the pollution. We can see pollution as a symptom of something that is wrong with the other factors. Eco-politics should preferably deal with the primary causes, not the symptoms.

Optimum population

Demographers remain divided over the possibility of setting an optimum population size, and how large it should be. Current proposals have failed to impress growth-oriented theorists. As we near the limit for demographic and economic growth, delaying a decision any longer will be untenable, say an increasing number of observers. It would be dangerous and irrational.

The definition of optimum population comprises three different groups of factors. The first concerns the population's preferred welfare standard to be achieved through its combined efforts; the second the population size; and the third all factors affecting population size and its achievement of the agreed standard of welfare.

Before we decide what an optimum population would be, we must agree on the standard of welfare the population seeks to bring about. The welfare standard can be given by a so-called welfare index or welfare function. This can be defined in several ways. Economists tend to define it by the monetary value of goods and services produced by the combined industry of the population. One speaks, for instance, of gross national product or its average, i.e. *per capita* income. Another definition of the welfare index uses indicators based on consumption or so-called welfare indices, social indices, indices on quality of life, etc. For my purposes, I shall assume in the following that the minimum standards of the UN's economic, social and cultural rights can be formulated such as to provide indicators for determining the standard of welfare.

The factors affecting population size and the population's realization of the desired standard of welfare include first nature, both renewable and non-renewable resources; second, real capital; third, accumulated knowledge, i.e. science and technology; and fourth, the social organization.

The most important production factor is the population itself. It constitutes the workforce. The question then is: how big must a population be to make best possible use of the other production factors when they collaborate on realizing the intended standard of welfare? The point being that the population can be neither too small nor too large without risking the outcome.

Against the background of the preceding factors, we can define, relatively basically, an optimal population as follows: assuming these production factors such as natural resources, real capital, science, technology and social organization remain constant, a population will be optimal when neither increasing nor decreasing it would improve the chances of realizing the anticipated standard of welfare.

In our case, the optimum population size, we think, would be the size facilitating the most satisfactory realization of welfare and fairness under N1 for as long a time as possible under the constraints set by natural resources. It will not be necessary in practice to assume that the optimum size is defined for a given time, a band rather, with an upper and lower limit. A more detailed specification of the band can be done through empirical trial and error.

As a guide in eco-political planning a norm of the following type has much to commend it:

> N3 Norm for resource and population management
> All decisions affecting resource and population policy should be made within the constraints of resource and population budgets which satisfy the criteria set out in main norm N2, and their preparation should adhere to the principles of rotational planning[12] which is based on systematically collected resource and population accounts.

Material welfare standards

So what would be the optimum human population size for planet Tellus? To answer this question we would have first to determine the welfare standard we wanted to achieve, and then the types and amounts of the resources we have at our disposal, requisite technology, capital, type of social organization, etc. This is no easy task, of course. I shall limit myself to addressing the welfare standard in this section and production factors in the next. I shall then return to the question of optimum population size.

Now with regard to welfare standards, we should first distinguish between the types of welfare requiring high consumption of natural resources and those that don't. There are several services and activities which play a key role in the good life and the good society which do not over-consume resources. This is true of cultural and social activities of many different kinds, including religious or philosophical contemplation, certain forms of research, art, music, cultivation of friendships, social care, outdoor pursuits, etc. Activities like these were more prominent in traditional societies. In modern, industrial-growth societies, they are constantly relegated to the back of the queue by activities fuelled by manufactured goods and consumption. As interest in a non-materialistic lifestyle grows, the less painful the transition from our present growth philosophy to a steady-state philosophy will be. And it will probably be felt as a change for the better, moreover.

Second, we should distinguish between goods manufactured from renewable and non-renewable resources. The production of food and fibre for clothing and houses, along with numerous other products, can be based on renewable resources and remain intensive even if the production of other goods based on non-renewable resources is reduced. It would be reasonable to expect the use of highly mechanized production forms to fall with increasing prevalence of labour-intensive, intermediate technologies. This would lead in turn to increase the proportion of the population working in primary industries, mainly farming, and in the non-mechanized secondary industries such as craft trades. The human factor would be strengthened in labour processes, increasing the preponderance of creative and meaningful jobs, which we must count as one of the most valuable welfare offshoots.

A crucial question, remains, however. What should be the level of consumption of goods produced from non-renewable resources, such as metals and fossil fuels?

If we argue on the basis of economic, social and cultural rights under N1, as well as the supplementary norms N1a–N1d, the answer would appear to be this: the consumption level should be as low as possible. In other words, things produced from non-renewable resources should be capital goods for use in the manufacture of other goods derived from renewable resources. We should generally seek to produce simple, reliable tools and implements leading to a certain improvement of the traditional technologies in agriculture and handicrafts. Things like iron spades, iron ploughs, scythes, saws, planes, etc., of use to thousands of coming generations, should be given priority ahead of machines for short-lived luxury consumption like private cars, jet planes or holiday trips by plane, speedboats with energy-guzzling engines, etc.

How can all this be justified? If we argue from N1a and N1d, the argument would likely go like this. Primarily, these resources should be divided equally among people living today and in the future. For instance, anyone running a farm should have the same claim on simple tools and implements of iron or steel. Second, we could consider whether to allow some people to increase their consumption beyond rightful minimum levels. The condition here would have to be that it was in the best interest of all implicated parties, including the least favoured parties today and tomorrow. These differences could be justified on the basis of labour distribution. It would be reasonable, for example, for the people who produce capital goods to enjoy a higher consumption quota of non-renewable resources than others.

As we have already discussed, there are gross inequalities in the consumption of non-renewable resources in the world today. In affluent countries, consumption levels lie high above minimum entitlements, and in the poor far below. One objective of a long-term resource policy should, in this light, be to close the gap between highest and lowest levels. They should meet at the minimum entitlement level. Why not higher? Because the non-renewable resources should last for as many generations as possible. The faster they are consumed per generation, the fewer generations will be able to enjoy them.

From these considerations we can derive the following norm:

> N4 Sub-norm for material welfare standard
> The private and public consumption of goods and services produced on the basis of non-renewable resources should be set as low as possible within the limits of what is acceptable under N1.

Resources policy principles

Given that the material welfare standard is determined in compliance with the principles presented in the last section, what would be the impact on resources policy? The following principle is pertinent:

> N5 Resources policy main norm
> Production of goods and services on the basis of natural resources should be maximally reproductive and minimally extractive.

If primary, secondary and tertiary industries adhered to this norm or something like it, it would have a very wide ranging impact indeed. I shall not attempt to explain what it would mean for agricultural and forestry policy, fisheries policy, fish and game management, energy policy, industrial policy, land use planning, etc.

I shall mention just the one point. It concerns relations between non-renewable and renewable resources. The exploitation of several types of non-renewable resources is likely to overtax renewable resources, especially the ecosystems. This is because of the extraction processes themselves and pollution. For instance, it is true of coal and metal ore extraction which is hugely damaging to the countryside; the same is true of oil extraction, especially the seabed, where there is the risk of harming marine and coastal biology. It is true of nuclear energy, with its potential to harm all manner of life through radioactive radiation, etc. A norm like N5 indicates the need to control extraction processes of this nature very stringently indeed. As far as Norway is concerned, it ought to mean not building nuclear power stations. It would also be wise not to extract North Sea oil.

A point of view like this obviously affects choice of technology: it obviously calls for material and energy-saving technologies:

> N6 Main norm for choosing technologies
> Production of goods and services on the basis of renewable and non-renewable resources should be based as far as possible on soft technologies.

What will be the impact of this on the organization of economic life or, more generally, the organization of society? The transition from traditional agrarian societies to modern industrial societies was also a transition from smaller to larger economic units. The industrial mode of production has so far entailed an increasingly specialized division of labour, and in turn has increased the rate at which goods and services are exchanged. All this needs a modern, effective means of communication, of course, which in turn requires the enormous consumption of non-renewable resources, especially metals and fossil energy sources. To realize a norm like N5, these resources would need to be severely rationed. It would necessarily entail downsizing the communication network, resulting in the downsizing of the economic units. Instead of the current gigantic level of global trade, we would have to build small societies, preferably steady-state societies producing mainly for the domestic market within the constraints set by the norms we have already mentioned. A move in the direction of steady-state societies would also mean a move in the direction of more self-sufficient societies in terms of goods and services production. But, as we have already mentioned, a certain exchange of industrial goods would be sensible, especially capital goods. The purpose of all this could be stated as the following principle:

> N7 Main norm for organizing economic units
> The economic units for production, exchange, distribution and consumption of goods and services should be as decentralized as possible under such constraints as conform with human rights norms under N1, the eco-political main norm under N2 and the other norms mentioned above.

One necessary condition for progressing towards a steady-state society is probably far greater self-governance at all levels of society, especially the lowest level. If people are powerless in the face of what is happening in their immediate environment, they will also probably feel ill-equipped to deal with issues further afield in time and space. The more control people have over their immediate conditions of life, the more motivated they will likely be to take responsibility for the more comprehensive, longer-term consequences of societal change. Compliance with N7 will probably reinforce this tendency.

Population policy principles

I return now to the issue of optimum population size in respect of the population of human beings on our planet. Where do the upper and lower limits go for an optimum population?

The question can obviously be answered in many ways. I shall limit myself here to outlining an argument based on the premises introduced above. One of these premises is that the earth's population should endure for as long as possible. Another is that all people in all generations have the same moral right to live well, at least to have their minimum rights met, as defined in the economic, social and cultural human rights, and therefore to enjoy a fair share of the natural resources used to realize this end. Concerning these resources, non-renewable resources are a scarcity factor. If non-renewable resources are to be shared equally among earlier and later generations, the earlier generations will have a moral duty to ration their annual consumption of these resources. This would necessitate in turn a lower material standard of living, simpler technologies and smaller population.

The question can be framed as follows: how many people can optimally live at an economic level which satisfies the minimum rights norms under N1 for the longest possible time if the production of goods and services is based largely on soft technologies, renewable resources and the contribution of non-renewable resources is heavily rationed?

It might help in answering this question to look at the size of the population around 1750. The earth's population at that point in time was mainly organized in agrarian societies, but there were also hunter and gatherer societies as well. These societies lived mainly on renewable resources. The population then counted around 600 million.

This figure might give us an idea of the size of an optimum population for the earth as a whole. Answering the question must obviously be rather speculative. On the one hand, we need to take factors enabling high production into account. They include conversion of land into new agricultural areas in large parts of the world over the last couple of centuries, and scientific and technological progress, such as the development of ecology as a field of study in its own right. There would be no question of returning to traditional agrarian technologies, but developing new, intermediate technologies, as discussed above.

On the other hand, however, there are constraining factors too. Much of the resource base has been depleted since the eighteenth century. Soil erosion is still accelerating, increasingly as a result of deforestation and overgrazing. A telling figure is the rate of desertification. Between 1882 and 1952, the percentage of the earth's dry land covered by deserts increased from 10 to 25 per cent. This trend could have been reversed by intensive reforestation. The rate of extinction of species of animals and plants is today many hundred times what it was in the Quaternary period, when it also was much higher on average than earlier periods in the earth's geological history. The main causes of this damage are growing populations and rising living standards. These factors press up against the limits of the earth's resource capacity, and the effect is exacerbated even more by pollution from the hard technologies used in the industrialized countries.

When we take these factors into consideration, the upper optimum population limit would appear to be at least no higher than the earth's current population. In fact, there are strong indications that it should be much lower. Moreover, no one says we have to fill the quota completely. There could be very good reasons for setting it as low as possible, that is, in the vicinity of the lower limit in the optimum band.

For the sake of argument, let us give these speculations some credence. The next question then will be the following: how long would the optimum population-reducing period last before we reached the desired level? In *Blueprint for Survival*, a couple of hundred years is suggested, with the addition of coordinated network downsizing plans. The idea is good, even if it would be difficult politically to carry out.

As a principle of population policy facilitating a steady-state society at an optimum level, the following norm may have something going for it:

> N8 Population policy main norm
> The population should be reduced to a size that makes it possible to satisfy human rights norms under N1 for as long as possible under the terms of the resource budgets determined under N3 and technologies determined under N6.

We have already mentioned the possibility of regulating population growth by increasing mortality rates and reducing birth rates. The former has been used in many earlier civilizations. The approach has been used in modern times as well, under Nazi rule, for instance. As long as we presume humanistic norms of the type entailed by human rights, these methods are out of the question. The following norm can be seen as a basic humanistic norm for any form of population control:

> N9 Whether population policy endeavours to achieve optimum population growth or zero growth at an optimum level, all positive population measures should seek to reduce the birth rate in relation to the death rate, and no positive measures should seek to raise the death rate relative to the birth rate.

Advocates of a fascist ideology have objected to the humanistic norms; together with other factors they are blamed for the current population explosion. This occurred, they reason, because improved food production and modern medical technology reduced the death rate. If humanistic considerations had not motivated organizations and governments to implement modern medical, hygienic and other preventative measures, the population explosion would never have happened. And what then is the result of these humanistic interventions? A child is saved from a fatal disease, but only so it can die instead of hunger or war. And anyway, you only increase the number of people exposed to these afflictions. Since the consequences are repellent, and contingent on humanistic norms, the norms must be repellent as well.

This argument, however, denies its own premises. What's wrong is not necessarily the humanist norms, but the way they are practised. Application has been too one-sided and short-sighted. The only aim has been to reduce the death rate; no steps have been taken to facilitate a slower birth rate. It may therefore be impossible to avoid giving the following population policy principle some weight:

> N10 Positive population measures aimed at reducing the death rate should not be enacted unless compensatory measures are enacted to reduce the birth rate such that population growth as a whole remains within the parameters of optimum negative growth or zero growth.

The norm would obviously need careful application. Its most important task is to highlight the importance of birth control. It must be given a far more central place than has been the case so far. For every practical implementation of these norms in every geographical region, we would have to estimate the annual birth rate for the whole population in the region. It could be estimated for all stages of the development, both for curbing positive growth and downsizing to the optimum level and for zero growth at this level. We shall not pursue prevention techniques, or legal, political, administrative and socio-moral problems attached to the implementation of these programmes. We limit ourselves to stressing the importance of these questions.

The Chinese may have found a promising way forward in this area. Their population policy consists of four elements, including high marital age and calculated family planning. Every married couple must, in 'consultation' with an expert, lay a plan for the number of children they would like, when they want to have them and how they intend to avoid having more children than planned. The system provides for a combination of highly decentralized decision making on the basis of general principles drawn up by the Chinese government after a comprehensive assessment of the population situation. The Chinese also make use of economic incentives, such as regressive child benefit and loss of certain types of support for couples who deviate from the agreed plan.

Putting the standpoints into practice

Supplementary normative and descriptive premises

The argumentation outlined above is based mainly on normative premises derived from the UN's economic, social and cultural human rights and descriptive premises derived from ecology and auxiliary disciplines. The argumentation points towards the eco-political main norm N2 favouring a steady-state society at a constant optimum level with regard to economy and population.

The next question concerns how we might or should attempt to translate this policy into practice. What sort of strategies and tactics have the best justification?

It will be easier to take a stand on this question if we allow for some additional normative and descriptive premises. Of the relevant sources of premises, the UN's civil and political human rights are particularly pertinent when it comes to normative premises, while the social sciences and lessons of everyday life are key when it comes to descriptive premises.

The civil and political rights, which I summarized in an earlier section, arguably define several different human freedoms. In important ways they constitute the backbone of the norms underlying democratic society. In an eco-political context, these norms could well clash with the preceding norms. Two main problems facing a democratic eco-policy can be framed as questions: to what degree is it possible to pursue these freedoms while effectively adhering to the resource and population-based guidelines outlined in the present text? And if the various freedoms and norms were to work at cross-purposes, which should we prefer? I shall restrict myself here to formulating the questions.

I have assumed in the above that the eco-political main norm N2, along with the other norms, could provide a suitable framework for an ecological and environmental movement embracing all political colours. They are, however, very general guidelines, and would have to be defined, specified and possibly modified in light of situations on the ground, such as campaigns or actions to solve different ecological conflicts. Since we are talking about extremely comprehensive practical problems, there is reason to believe that the strategy would need to be of similar complexity. Parliamentary and non-parliamentary means will be required. In concluding I want to air some ideas of a highly incomplete nature in relation to this problem.

Parliamentary means

If the population of all human beings is to achieve something like a steady-state society at an optimal level, all human sub-populations would have to conduct themselves under the constraints set by the norms defining the steady-state society. The steady-state society can possibly be defined by a system of norms of production, exchange, distribution and consumption of goods and services and of the reproduction of the species. Most legal and socio-moral norms relate to these matters directly or indirectly. This type of norm system could be designed on the principles of rotational planning with the help of resources and population budgets adapted to regional and local conditions and adjusted in light of new information. Democratic decision making procedures could be included in this form of planning at several junctures, both when determining the overall parameters for ensuring the overall system's macro-stability, and when decisions are taken within these parameters and which ensure micro-viability in all areas of life.

It would necessitate changes to the norm systems which essentially underpin behaviour and interaction today. These norm systems would have to be integrated in compliance with the standards of the steady-state society. It would entail an enormous process of change for all organizations and most individuals at the various international, regional, national and local levels. It would entail changing inapplicable norms currently institutionalized within most of these organizations, both public and private. And the mindset of individuals that had internalized (accepted) these inappropriate norms would also have to change.

A highly complex eco-strategy could conceivably invoke the following 'parliamentary' argumentation strategy. We have already mentioned the wide range of organizations that have institutionalized norms of the type included in the UN's human rights. Obvious examples are the different UN organizations. For example, the purpose of the International Labour Organization is to address conditions of work; the Food and Agriculture Organization attends to food production; UNCTAD sees to education and culture, etc. A similar division of responsibilities is found in governments and public administrations. It is emulated, for example, by political parties, unions, idealist organizations of many different types, etc.

To a greater or lesser extent these normative premises are already accepted by members of these organizations. An argumentative strategy which draws on these factors could proceed by the following three steps: (1) those employing the strategy could help to assemble the relevant normative premises already established in the particularly organization, they could highlight them, give them a more precise wording, appeal to them, etc.; (2) they could then enrich the norms by adding the descriptive premises of relevance to the different issues surrounding resources and population questions; (3) they could then provide assistance by drawing the soundest practical, eco-political conclusions in light of these premises. These conclusions, I presume, would support N2 in the majority of cases. What still needs to be done today is possibly to enact steps 2 and 3 of the strategy. By intensifying efforts on these two steps, it should be possible to accelerate the necessary changes. For those with an interest in pursuing the matter, many tasks remain to be tackled.

Non-parliamentary means

The number of non-parliamentarian means of persuasion is legion, including non-violent and violent means. I shall limit my remarks to the former. The most effective non-violent means of persuasion is that deployed in combination with parliamentary methods. One of the basic strategic models in this area is the following:

> Those in authority in any organisation, public or private, depend on support to exercise their authority. They need the support of other people and groups within and beyond the confines of the organization. This support, however, is fragile and is likely to crumble in the face of non-violent opposition.

> The non-violent methods entail different degrees of non-cooperation with persons in authority.[13]

If persuading the leadership of a given organization to adopt a given policy is on the cards, we should first perform an analysis of their power basis and identify the groups whose support they rely on and work together with. This defines the strategic parameters of the campaign. The tactical forays will then consist of encouraging precisely these groups to withdraw their support and willingness to cooperate, unless the leadership gives way and accepts the demands. I shall not pursue the diversity of campaign schemes of use in these tactical operations. In general, the better the preparatory strategy analysis, the better the chances of tactical success.

There are possibilities in this field to combine ecological policy framework plans and local democracy. Coordinating framework plans often leads, unfortunately, to centralized decision making. Implementation, on the other hand, relies largely on local initiatives and special interest group advocacy in a range of areas. It is a vibrant area for local partnerships with ecological and environmental groups working at the grassroots level and upwards.

Notes

1 Cf. Ch. Perelman, *De la Justice*, Brussels, 1945, translated as 'Concerning Justice', and included in Ch. Perelman, *The Idea of Justice and the Problem of Argument*, London, 1963.
2 J. Rawls, *A Theory of Justice*, Cambridge, MA, 1971, esp. pp. 60–5, 302–3.
3 Cf. G. Myrdal, *Beyond the Welfare State, Economic Planning and its International Implications*, Yale, 1960 edn, New York, 1967, p. 147.
4 Cf. A. Næss, *Økologi, samfunn og livsstil. Utkast til en økosofi*, Oslo, 1974, p. 171f., 'Likestilling i biosfæren. Økosofi T'.
5 Cf. E.F. Schumacher, 'Western Europe's Energy Crisis', *Resurgence*, vol. 5, no. 4, 1974, p. 11.
6 The typology was devised by Robert Clarke at UNESCO's science department and cited by R. Jungk, 'Politics and Technocracy', *The Spokesman*, 23, 1972, pp. 8–9.
7 The following articles by H.E. Daly are highly recommendable on the theory of steady state society: 'Toward a Stationary-State Economy', in J. Harte and R.H. Socolow (eds), *Patient Earth*, New York, 1971, pp. 226–44; 'The Stationary-State Economy', *The Ecologist*, vol. 2, no. 7, 1972, pp. 4–12; 'How to Stabilise the Economy', *The Ecologist*, vol. 3, no. 3, 1973, pp. 90–7.
8 See for example E.P. Odum, *Fundamentals of Ecology*. New York: Saunders, 1971, chapter 9, 'Development and Evolution of the Ecosystem', esp. p. 264, 'Concept of the Climax'.
9 I can recommend in this connection S. Kvaløy, *Økofilosofisk fragment Nr. IV*, SIFUA, Tr.h., 1974.
10 See E.P. Odum, ibid., pp. 166–70, and D.H. Meadows, D.L. Meadows, J. Randers, and W. W. Behrens III, *The Limits to Growth*. Washington, D.C.: Potomac Association, 1972, pp. 34–44.
11 K.E. Boulding, 'The Economics of the Coming Spaceship Earth', in *Environmental Quality in a Growing Economy* (ed.) H. Jarrett, Baltimore, 1966.
12 Rotational planning means revising shorter and longer term plans at certain intervals.
13 G. Sharp, *The Politics of Nonviolent Action*, Boston, 1973. See also A. Næss and J. Galtung, *Gandhis politiske etikk*, Oslo, 1955.

12

PRECAUTIONARY SCIENCE

The ecophilosophical foundation

Karl Georg Høyer

The precautionary principle – PP – is a trueborn child of the late modern ecological and environmental discourse. It is a discourse that was initiated in the early 1960s, where the book *Silent Spring* (1962) by the American biologist Rachel Carson is given the due credit. Carson describes the disastrous ecological effects of the extensive use of chemical insecticides, especially DDT, in industrialized agricultural practices. The effects were a spring where birds were not singing and insects not humming – a silent spring. She devoted her book to Albert Schweitzer and his claim: *'Humans have lost their ability to foresee and prevent the foreseen to happen. They will end by destroying the earth.'* They were both forerunners. To foresee and to prevent are both basic aspects of the precautionary principle.

Already around the 1970s this particular discourse made its first institutional tracks. In 1969 the United States got its first fundamental environmental legislation requiring impact assessments in order to improve its abilities to foresee and prevent. The first UN global environmental conference was set up in Stockholm in 1972, where key environmental principles were first officially coined in a broader context. Two important principles were: *'the reverse burden of proof'* and *'doubt shall be to the benefit of the accused – nature'*. The first implies that it is the obligation for the responsible ones behind an encroachment in nature to 'prove' that it is safe; it is not the responsibility for the opponents, environmental organizations, public authorities, or others. The second implies that when there are doubts to what extent an encroachment in nature has detrimental effects, the doubt shall be to the benefit of nature. Both are principles further contributing to the development of the precautionary principle.

A conceptual framework

The PP is subjected to various interpretations, in particular in relation to three basic aspects: *prevention*, *uncertainty* and *damage characteristics* (Høyer 1999). Both in terms of its conceptual core and policy implications, the PP is a *preventive* environmental-political principle. It hinges on the fact that precaution implies the prevention of damage, rather than reparation afterwards. However, this does not imply that all types of preventive environmental policy and measures are precautionary approaches. They may be non-precautionary or precautionary to a varying degree. On the other hand, a policy or measure implemented based on the PP can never be non-preventive. This implies that we shall use the terms *precautionary policies and measures* and *not preventive* when they are based on the PP; a distinction which should be further clarified in relation to the aspect of uncertainty.

All official formulations of the PP hinge on the aspect of *uncertainty*; 'lack of full scientific certainty should not be used as a reason for postponing measures to prevent environmental degradation'. We shall later elaborate on various interpretations of the term 'uncertainty'. The fact that such a lack exists – or that uncertainty exists – is a *sufficient* condition for implementing precautionary policies and measures in order to avoid the environmental degradation. The implication is that the PP *does not apply* when it is certain that environmental damage will occur. Nor does it apply when we are fairly certain as to the occurrence of such damage. In such cases implementation of policies and measures must take place on another basis, for instance as part of a more general *preventive* environmental policy.

Damage characteristics is a third aspect; 'where there are threats of serious or irreversible damage' and 'when there is a threat of substantial reduction or loss of biological diversity'. All such official formulations refer to serious or irreversible environmental damages. Even though this necessarily is a matter of interpretation and dispute, it underlines that the PP does *not* apply to *any form* of environmental damage. There must be possible serious or irreversible damages and uncertainty as to their existence and extent and/or possibility of occurrence. With a basis in known cause–effect relations there may be a possibility for more serious damages than we know of; and there may be a more fundamental lack of knowledge concerning the whole complex of cause–effect relations, but where the potential for serious or irreversible damage nevertheless is present. Another element of damage characteristics concerns *the type* of damage. Where the PP has been included in international treaties and declarations during and after the Rio conference they generally refer to *environmental damage*. However, it seems to be internationally accepted that the PP also applies to *human health damage*, even in those cases where the links to processes in nature are almost non-existent or of minor importance. This is not the least confirmed by several later EU documents and recommendations (Harremoës *et al.* 2001; EU 2000; Stirling *et al.* 1999). But it is also reflected in two later Norwegian public committee reports, one on health effects of xenotransplantation (Gjørv Commission, NOU 2001: 18) and the other on health effects of genetically modified food products (Walløe Commission, NOU 2000: 29). Both have lengthy and explicit discussions of the PP, and give concrete recommendations as to its application.

A further qualification is that the PP applies as much to *uncertainties over agents* as to those over effects and possibilities. Established regulatory systems (Risk

Assessment and Environmental Impact Assessment) represent prospective appraisal. Uncertainties over agents do, however, require *retrospective appraisal*; which starts with the documenting of fairly manifest health or environmental effects and seeks the identification of cause–effect relations and possible agents. Even though there may be uncertainties about the grounds for the follow-up targeting of a particular agent, the PP may still be legitimately invoked to defend continued action on this agent, until such uncertainties are resolved (Harremoës *et al.* 2001).

In policy applications of the PP there is a need to differentiate between the following aspects (Deville and Harding 1997): (1) Does the PP come to application? (2) How cautious should we be? (3) What precautionary measures can be applied? (4) What precautionary measures should be applied? All aspects are included in a concept of 'policy application'. There is, however, much confusion in this context. Some seem to imply that application of the PP always means absolute prohibition or at least a moratorium. This is not necessarily so. There is no one absolute right strategy to follow or measure to apply in a straightforward sense, once we have decided that the PP comes to application. Actually there will be a multitude of possible strategies and measures (Kaiser 2001). These are some generic examples of *direct precautionary measures* (Deville and Harding 1997):

- to stop the activity or modify it to significantly lessen possible impacts;
- to reduce the uncertainties, so potential impacts can be re-evaluated;
- to carry out a broad assessment of alternatives, and to opt for an alternative with substantially less impacts;
- to use rigorous quality systems to limit impacts during operations;
- to monitor changes and impacts closely, so that the activity may be halted before damage becomes serious or irreversible.

In addition there is a multitude of possible *indirect and institutional measures*, to facilitate the avoidance of threats in the longer term by encouraging processes and actions that produce precautionary results.

PP in international environmental politics

In this form the PP has since the mid 1980s become an integral part of numerous international environmental declarations, treaties and conventions, besides the integration into a multitude of national legislations. It was initiated in 1984 with the first international convention on the protection of the North Sea against pollution, and not the least strongly emphasized in its 1987 London declaration: 'in order to protect the North Sea against possible damaging effects of the most dangerous substances it is necessary to apply a precautionary approach, which may imply measures to stop the emissions of such substances even before causal effects have been established based on certain scientific proofs.' This was the same year as the launching of the UN Brundtland Commission report on sustainable development, however without any explicit references to the PP. This should change very soon. One of the main UN-regional conferences to follow up the Brundtland Commission report was arranged in Bergen in 1990. It was part of the preparation for the first UN global summit on sustainable development to be held in Rio de Janeiro in 1992. In the common ministerial declaration from the Bergen conference an absolute link between sustainable development and PP was established, and the PP was defined as 'lack of complete scientific proof must not be used as an excuse for not implementing measures avoid environmental damage'. In the same wording the PP was included in the main UN declaration from the 1992 Rio conference, as well as in the two crucial conference conventions on climate change and biodiversity. In other international processes the EU was a forerunner in including the PP in its legislative systems. It was thus included in the preamble of the Maastricht Treaty, and later on in numerous EU declarations and treaties, as well in their separate environmental directives, among other as regards application of genetic technology. In this area the particular Norwegian legislation from 1992 emphasizes that the PP shall come to application through this wording: 'A reasonable degree of doubt about the risks for damages shall at the outset imply that concessions are not given.' Actually in this context there were explicit references to the long traditions in Norwegian ecophilosophy. Later on, in 1997, when discussing the future politics of sustainable development, an almost unanimous Norwegian Parliament decided that the PP should be one of two main principles for the development of a sustainable environmental politics.

The risk society

The PP has a rooting also in *sociological theory* on the *risk society*, as it originally was presented by the German Ulrich Beck in his 1986 book on the *Risk Society* (Beck 1986). He (Beck 1992, 1997) points out how what he calls *modernization risks* emerge both local-specifically and universal-non-specifically, and how unreliable and unpredictable are the mysterious ways of the damage effects. Consequently, in modernization risks relations that are separated substantially, spatially and temporally are linked causally with and put in a common social and judicial responsibility context. Furthermore, he points out (1992) that in the classic industrial society science on the one hand is institutionalized, and thus the methodological doubt. However, the basis for and the impacts of scientific work are screened from the internally produced scepticism. With the risk society the doubt is extended to the basis for and the risks implied in scientific work, at the same time with the consequence that science is generalized and demystified. Beck has, however, no mentioning of the PP. This is actually rather a shortcoming in all his writing as application of PP both as a matter of politics and of science turn up as rather evident implications of his risk society. But he comes very close to really arguing for such a principle (Beck 1997: 286):

> Technological (or Ecological) democracy is my Utopian idea for a responsible modern. I envisage a society where the consequences of technological

> development and economical change are debated *before* the important decisions are taken. The burden of proof as regards future risks and dangers, included the current environmental damages, must be the responsibility of the actors behind the troublesome actions, and not the responsibility of the victims: from the principle of the polluter *pays* to the principle of the polluter *proves*. At last there is the need to establish a new set of standards for proofs, correctness, truth and unity in science and law. What we need is thereby no less than *another enlightenment* which opens our minds, eyes and institutions for the dangers inherently produced by the industrial civilization. (My translation)

Included in this discourse on 'risk society' and 'social theories of risk' is the concept of *post-normal science* as it has been elaborated by Silvio Funtowicz and Jerome Ravetz (1992). This is about the requirements for a science that relates to cases where stakes are high and uncertainties are large. It is, among others, articulated as a shift from the 'normal science' understanding of 'hard facts and soft values' to the post-normal 'soft facts and hard values'. The 'normal' way of scientific development puts weight on reducing *internal* risks and tends to suppress information on external risks towards society. The alternative approach is to promote a science that encompasses both internal and external risks, and accepts the constituting situation of 'soft facts and hard values'. These are some key elements of my term *Precautionary Science*, which I shall return to. I prefer this term to post-normal, as too much of the later scientific discourse in this field is dominated by postmodernist and relativist contributions.

The ethical foundation of PP

The German philosopher Hans Jonas outlines in his book *Das Prinzip Verantwortung* ('The Imperative of Responsibility', 1984) an ethical foundation of PP. Ulrich Beck (1999) emphasizes that his risk society theory is extensively inspired by Jonas and his moral philosophy and responsibility ethics. But neither Beck nor Jonas give any explicit references to PP. In the outline below I'm very much indebted to a work by the Norwegian philosopher – and former environmental activist – Alfred Fidjestøl (2004) about Hans Jonas and his ethical theories.

His *responsibility ethics* is in a real sense a consequence ethics. It is not the good intention or thought that counts. The actions that interests Jonas are often based on the best intentions, but they still have the worst of consequences. The sum total of all actions with good intentions is threatening the very conditions for all future human existence. A problem however is that, as emphasized by Ulrich Beck, the consequences of our actions in the technological society are not any more obvious to us. In many cases they can neither be seen, smelled nor tasted, and neither do they always hurt. Beck (1986) illustrates this by emphasizing that a major problem with radioactivity is that it doesn't itch, and a key task actually is to make it itch. According to Jonas this implies that we have a duty to get knowledge about the possible long-term effects of our collective actions; a duty of knowledge. Neither individually, nor collectively, is it acceptable to get away with 'we didn't know'.

On a societal level there is according to Jonas a demand for science to come up with the best knowledge about possible consequences of the various human actions, actually necessitating *interdisciplinary* work. Jonas thereby claims a clear link between responsibility ethics and interdisciplinarity. This knowledge can obviously not be available to everybody individually; it presupposes scientific data, methods and theories. The ethical situation is that what is morally right or wrong is not solely a matter to be decided by individual conscience and the individual's inner conversation with himself/herself. Good and right can then only come up through a interplay with science.

However, only naive positivism can claim that it ever is possible to get complete and certain knowledge about all possible future consequences of our collective actions. The divide between what we can proclaim for the future and what we most certainly can carry out today will never disappear. If the opposite claim is the basis for our current actions, then the future will continue to surprise us. The knowledge duty in the responsibility ethics is thereby only meaningful together with a duty to handle the uncertainties, which according to Jonas requires some principles. My emphasis is that PP, both in politics and in science, is such a principle. And Jonas comes indeed very close to this. He emphasizes that we must base our actions on the most negative possible consequences; in the PP discourse termed *'worst case scenarios'*. This is because if the worst case scenarios should be real after all, the consequences are vastly more serious than if it in due time proved that our actions for avoidance were unnecessary. Even a relatively low risk can make the most radical countermeasures reasonable if the possible, but not probable, long-term consequences are serious enough. Jonas emphasizes the necessity to handle the uncertain as if it were certain.

We shall accordingly behave in the opposite way to Descartes. He set forth to doubt about everything, even his own existence, and came up with his own doubt as the only thing he could not principally doubt. Descartes could only absolutely trust that there is a doubting *I*. In the responsibility ethics we must on the contrary trust that which is radically uncertain. Jonas emphasizes that we thus must carry through radical political and societal changes based on the possible consequences which there certainly can be raised doubt about. A precautionary science here represents no less a radical change from the Cartesian legacy, as it also implies to handle the most uncertain as if it is certain, actually *the uncertain beyond doubt.*

With obvious connotations to the Kantian categorical imperative Jonas puts forward a moral imperative: *'Act in such ways that the consequences of your acts are consistent with the continuation of real human life on earth.'* He still considers it to be very different from the individualistic Kantian approach. It is an imperative that to a much larger extent is directed towards the collective and political levels, and towards the future more than today, where sacrificing actions today may be necessary to save the future generations. As grounds for motivating to such actions he emphasizes *fear*, while claiming that both *fear* and *the evil* traditionally are hidden categories in the history of philosophy; ethics has too much focused on the good. He thus raises the issue of a particular *'fear-heuristic'*, where just the fear for apocalyptic consequences of our collective actions shall be the help human societies need to motivate necessary changes; it is only when the human societies consider themselves really threatened that it gives meaning in claiming a duty for future human existence.

The extended ecophilosophical foundation

The works by Ulrich Beck on the risk society are mostly anthropocentric. In this he belongs to a long tradition in sociology, ever since Durkheim and Weber. My representation of the Hans Jonas ethics is also anthropocentric. But this is not fair to all of his philosophical works. He is one of few moral philosophers who really has made in-depth explorations of the importance of other species and their relations to humans understood as a species belonging to nature. His much renowned book from 1963 is thus entitled *The Phenomenon of Life: Toward a Philosophical Biology* (Jonas 1963). The development of his responsibility ethics published in the 1984 book is actually very much based on this earlier work.

Non-anthropocentrism is an important part of Nordic ecophilosophy, and has actually strongly contributed to the wider international discourse on environmental or ecological ethics. I shall here include some of this in order to present an ethically necessary extended ecophilosophical base for the PP. Again it is most basically a *responsibility ethics*, largely in line with Hans Jonas. Our responsibilities, however, are not delimited to human existence, but are including the existence of other species and nature as a whole. It is not only the existence of humans that is threatened, but the existence of other species and the ecology they are depending on. Other species – and the ecological systems they are part of – are considered to have moral value, while humans of course are the only species with moral capacity as well. Arne Næss – in this book – terms this the *inherent value* both of singular species and their environment.

Georg Henrik von Wright is one among the Nordic ecophilosophers who has stressed the importance of an extended responsibility ethics. He (von Wright 1993) emphasizes that such an ethics should address:

- our responsibility towards future generations of humans;
- our responsibility towards animals, which are our nearest of kin;
- our responsibility towards other nature;
- our responsibility to our own natural bodies.

Related to my own works on understanding and defining the concept of *sustainable development* I have pinpointed the following fourfold aspects of the extended ethical responsibility (Høyer 1999):

- our responsibility towards intra-generations of humans and their bodies;
- our responsibility towards inter-generations of humans and their bodies;

- our responsibility towards intra-generations of nature and its species;
- our responsibility towards inter-generations of nature and its species.

Basically it is a matter of both *intra-generational and inter-generational responsibility* towards humans, their natural bodies, the other nature and its species. In this there are several extensions of the Hans Jonas ethics. First, our responsibilities towards our own human bodies are emphasized, in reality making restrictions on the development of gene technology for human purposes. Second, the responsibility towards humans is not only a matter of the future. The current – intra-generational – situation is just as crucial. Humans – and human societies – far away from where the actions take place are most often innocently hit by their consequences. Actually most of the human existences today are innocent victims of the actions by a few, which of course is not acceptable. And, third, also nature and its species are similar innocent victims of our actions, now – intra-generationally – and in the future – inter-generationally.

This is where the PP comes in. It applies just as much to humans and their bodies as to the other nature and its species. And it does not only apply to uncertain future situations, but just as much to the uncertainties in the world of today, the far away humans and the both nearby and far away nature and its species. The PP applies in all cases where there are uncertainties about the risk and seriousness of actions with possible consequences for one or many of the four responsibilities above. With its application follows measures of today to counteract the even vague possibility of the uncertainties to become real.

A *precautionary science* responding to this is very different from normal science. It covers the following efforts:

- to treat uncertainties of possible consequences as temporary certainties;
- to deliberately research for such uncertainties;
- to engage in interdisciplinary research in order to strengthen the focus and knowledge on uncertainties;
- to communicate and highlight such uncertainties within the research community;
- to warn the society about possible adverse consequences from the research you participate in, even when there are large uncertainties to this knowledge.

The two last points are actually included in the Norwegian 2007 *research-ethical guidelines* for all natural and technological sciences. The guidelines have made the PP obligatory for all Norwegian scientists and research in these fields. In principle a real scientific breakthrough, but with such far reaching implications that the real effects are still to be awaited. However, this is by far from taking place in Norway by accident; the long history and influence of the Norwegian and Nordic ecophilosophy as it is outlined in this book can easily be traced.

References

Beck, U. (1986): *Risikogesellschaft – Auf dem Weg in eine andere Moderne*. Frankfurt am Main: Suhrkamp Verlag.

Beck, U. (1992): *Risk Society. Towards a New Modernity*. London: SAGE Publ.

Beck, U. (1997): *Risiko og Frihet* (Risk and Freedom). Bergen: Fagbokforlaget.

Beck, U. (1999): *World Risk Society*. London: Polity Press.

Carson, R. (1962): *Den Tause Våren* (Silent Spring). Oslo: Tiden Forlag.

Deville, A. and R. Harding (1997): *Applying the Precautionary Principle*. Sydney: The Federation Press.

EU (2000): *EU Communication on the Precautionary Principle*. Communication from the Commission. COM (2000) 1. Brussels 02.02.2000.

Fidjestøl, A. (2004): *Hans Jonas. En introduksjon* (Hans Jonas. An Introduction). Oslo: Universitetsforlaget.

Funtowicz, S. and J. Ravetz (1992): 'Three Types of Risk Assessment and the Emergence of Post-Normal Science'. In: Krimsky, S. and D. Golding (eds): *Social Theories of Risk*. Westport: Praeger.

Harremoës, P. *et al.* (eds, 2001): *Late Lessons from Early Warnings: The Precautionary Principle*. Environmental Issue Report no. 22. Copenhagen: European Environment Agency.

Høyer, K.G. (1999): *Sustainable Mobility. The Concept and its Implications*. PhD thesis. Roskilde: Roskilde Universiy Center.

Jonas, H. (1963): *The Phenomenon of Life: Toward a Philosophical Biology*. New York: Harper and Row.

Jonas, H. (1984): *The Imperative of Responsibility: In Search of an Ethics for the Technological Age*. Chicago: Chicago University Press.

Jonas, H. (1993): *Das Prinzip Verantwortung. Versuch einer Ethik für die Technologishe Zivilisation*. Frankfurt am Main: Suhrkamp Verlag.

Kaiser, M. (2001): *Ethics, Science and Precaution*. Lecture notes. Oslo: NENT.

NOU (2000): Norwegian Public Commission Report. The Walløe Commission.

NOU (2001): Norwegian Public Commission Report. The Gjørv Commission.

Stirling, A. *et al.* (1999): *On Science and Precaution in the Management of Technological Risk*. Final report of the ESTO network study. Brussels: EC Forward Studies Unit.

Von Wright, G.H. (1993): *Myten om Framskrittet* (The Myth of Progress). Oslo.

13

NORDIC ECOPHILOSOPHY AND CRITICAL REALISM

Trond Gansmo Jakobsen

Nordic ecophilosophy is seen as the opposite to reductionism. In its very roots it is criticizing physicalism and materialism as developed in the Cartesian worldview. In this regard Nordic ecophilosophy is anti-reductionist and holist and represents non-anthropocentric as opposed to anthropocentric approaches. However, my main aim in this chapter is to show that even if Nordic ecophilosophy cannot be regarded as reductionist in the classical physicalist sense, it nevertheless represents a tendency to reductionism in a different sense. As classical (empirical) realism mostly reduces to physics Nordic ecophilosophy has a tendency to reduce things to ecology. What we are talking about is a form of ecologism, not physicalism. My focus in this chapter is to develop what I mean by Nordic ecophilosophy being reductionist in this sense and to say something about its implications. By then showing how critical realism can widen our perspective to an integrative pluralism involving a stratified and layered reality with powers, dispositions, emergence and change in open systems and on different organizational levels, from physics, to biology and ecology, humans and society, I want to open up space for a non-reductionist and non-ecologistic richer and deeper ontology as well as our understanding of our own place in it. This reflection involves dialectical critical realism as well as basic critical realism. By providing a brief introduction to metaReality one of the intentions of this chapter is to show to what extent critical realism might improve on Nordic ecophilosophical projects.

Introduction

What I am going to argue in this chapter might come as a surprise. I am going to say that Nordic ecophilosophy has a tendency towards reductionism. What I am referring to as Nordic ecophilosophy in this connection is Sigmund Kvaløy Setereng's and Arne Næss' work. As the term 'reductionism' can give unfortunate associations, a qualification needs to be made. I am not out to say that Nordic ecophilosophy is reductionist in any classical mechanist way, this would clearly be absurd as NE in its very roots are holist and eco-centered. Nevertheless, I do think it is reductionist in a specific sense, which I am going to explain. I must confess that my main point in introducing the concept 'reductionist' applied to NE is not to devalue or diminish the ecophilosophical contribution made by NE because I think that is both groundbreaking and important. It is rather to be able to clarify to what extent critical realist approaches can be seen as improving, deepening and transcending NE projects. Hence, my real reason for using the concept 'reductionism' in relation to NE is to show how critical realism gives us a more comprehensive and inclusive perspective. To give meaning to my provocative claim I want to explain what it amounts to. How can a philosophy that is in its very roots anti-reductionist nevertheless be seen as reductionist in a particular way? This needs explanation.

Traditional reductionism

First of all what I do *not* mean by reductionism is traditional reductionism. Traditional reductionism depends upon a view put forward by Descartes. According to Descartes every complex system or everything in the world from the human body, to animals, plants and congregation of matter can be explained by the parts of which it consists. That is to say to understand such a complex object as the human body or an animal's behaviour, there is nothing but matter in motion. Hence complex systems to be explained and understood must be reduced to their component parts. Just as a watch might be explained by its component parts the same is the case for any other complex living or non-living system.

Following Descartes' line of reasoning E.O. Wilson states that reductionism is the primary and essential activity of modern science.

> To make any progress scientists must mediate on the networks of cause and effect across adjacent levels of organization – from subatomic particles to atoms, say, or organisms to species . . . Its strong form is total consilience, which holds that nature is organized by simple universal laws of physics to which all other laws and principles can eventually be reduced. This transcendental world view is the light and way for many scientific materialists (I admit to being among them).
>
> *(Wilson 1998: 55)*

Descartes was reluctant to argue that reductionism could apply 'tout court'. As the human mind consists of different 'stuff' from that of matter he also claimed that human psychology could not be explained in reductive terms. Wilson is not so careful. According to Wilson 'universal materialism' is the most important question 'that can be asked in philosophy and religion' (1998: 54–5). So there is in principle no limit to reductionism:

> The transition from purely phenomenological to fundamental theory in sociology must await a full, neuronal explanation of the human brain. Only

> when the machinery can be torn down on paper at the level of the cell and put together again will the properties of emotion and ethical judgment become clear.
>
> *(Plumwood 1993: 121)*

Obviously I do not mean that Nordic ecophilosophy is ruled by the type of reductionism above. Such a reductionism would make most ecophilosophical approaches rather uninteresting. If nature is understood as matter in motion or atoms in the void then there would be nothing 'out there' to be morally concerned about or to have any ethically responsibility towards. At most it would enable us to say something similar to William Baxter:

> My criteria are oriented to people, not penguins. Damage to penguins, or sugar pines, or geological marvels is, without more, simply irrelevant. One must go further, by my criteria, and say: Penguins are important because people enjoy seeing them walk about rocks . . . I have no interest in preserving penguins for their own sake.
>
> *(Des Jardins 1992: 49)*

No doubt, NE is in real opposition to such an anthropocentrism seeing nature and natural entities as outside the scope of moral concern. However, before we try to clarify what we mean by NE being reductionist then let's look at what alternative reductionist approaches might look like.

Social constructivism

The Edinburgh school (Barry Barnes and David Bloor) paid attention to how social factors not only organize and institutionalize science, but in a much more direct sense influence its knowledge content. Hence, social factors came to be seen as epistemologically important. As a consequence of this the Edinburgh school criticized traditional views of science as too asymmetrical in the evaluation of good and bad science. Successful science was seen as a way to truth, bad science was normally held to be such because there were some psychological or social factors that disturbed the scientific approach (e.g. of Lamarckism in the Soviet Union). Barnes and Bloor argued that all science needs to be studied in its social context. Hence, they also launched a symmetric model as an alternative to the classical asymmetric model. This is normally named the 'strong program' in science studies. It centered on the view that social explanation was to be foregrounded in science studies. In this regard earlier views of science were too abstract and too occupied with internal considerations – for instance, the idea that there is just scientific method and empirical evidence. Social studies of science became very popular in the late 1970s and in the 1980s. It was seen as important to get into the laboratories and reveal how scientists actually do research or what they are doing in their practice. It was important to pay attention to how social and experimental activities were directly influencing the knowledge and the results scientists were proclaiming. Andrew Pickering's book *Constructing Quarks* (1984) and Latour and Woolgar's bok *Laboratory Life: The Social Construction of Scientific Facts* (1979) are ethnographic studies of laboratory life. Latour's and Pickering's work are normally seen as involving a 'social constructivist' view of science and knowledge.

One main point made in social constructivist approaches is best understood in terms of the views developed by post-structuralist and postmodernist philosophers. In opposition to any classical approach to science and knowledge these thinkers were undermining the very idea that there is a way to 'mirror' reality or that we have any real access to reality in any classical empiricist/positivist or Cartesian way. We do not have any direct access to any real world or to social facts or human nature as such. To talk about the real world, society or humans is just another myth-governed narrative or fairytale. This is because postmodernism denies all metaphysical, ontological as well as epistemological important questions to be answered by anything else than human ongoing 'discourses'. It is through our discourses, ideas, theories and concepts that we shape the way the world is or what humans and society are. Thus natural things exist only as far as they are conceptualized or theoretized about. One startling example on how such a 'social constructivism' might work has been developed by Keith Tester. In his book on the practice of animal rights he states: 'A fish is only a fish if it is socially classified as fish, and that classification is only concerned with fish to the extent that scaly things living in the sea help society define itself' (Sutton 2007: 26). Thus a discourse idealist will say that fish is not only a fish, they are social or discourse constructed entities and they are different things according to how different discourses or societies define them. Fish is seen as food. Fishing is a sport enjoyed by millions where the main aim is not to eat but to compete. People keep fish in decorative tanks in their living rooms as pets; others are kept in hospital visiting rooms as ornaments that help people relax. Yet others see fish as creatures with rights that have to be respected, and they want both fishing and pet-keeping banned. There is no truth to the matter of what fish are. So strict social constructivists says that fish are not fish; they are in fact social constructions and they are constructed in different ways. And if fish is socially constructed, then the rest of the world or nature is too (2007: 27).

Surely such a postmodern or social constructivist approach would not make our ecophilosophical aspirations come out much better than in the reductionist/physicalist approach (see Baxter). The reason is that 'nature' as we know it would vanish from our pictures of things, it would be a completely subjective or cultural category. Something that would make us even more 'detached' from nature than is the case in the Cartesian picture. In postmodernism and social constructivist approaches it all depends upon the way we 'talk about talk'.

Preliminary conclusion

One aspect of the understanding of science and knowledge has been the view that science can detect reality as it is. To conform to such a view Cartesianism has been a driving force. In this regard reductionism is the view that all of the sciences reduces to physics. Hence this type of reductionism is also called physicalism, i.e. the view that at the 'end of the day' everything can be understood in terms of the physical sciences. The logical positivist thesis of the 'unity of science', Quine's thesis of 'ultimate science' and the early Putnam/Oppenheim thesis of 'metaphysical realism', imply such a reductionism and physicalism. However, following our outline of social constructivist approaches we are now able to see the 'contours' of an alternative reductionist approach. In this case we are not talking about reducing everything into physics, or to matter in motion or whatever. What we are talking about is reduction to social science or sociology. This is to say that science and knowledge is nothing but a social phenomenon, i.e. a phenomenon socially constructed by us. There is no world 'out there' everything is 'in here'. It is all part of a 'cultural game' and nothing but. If we are to understand science and knowledge we investigate the scientific community or 'social discourse'. The reductionism dealt with in this connection is that 'ontology' reduces to sociology. This position is normally labeled 'the sociology of knowledge thesis'.

However, dealing with reductionism in this latter sense there are different types of reductionism reflecting our favourite science. There is at least physicalism, biologism, psychologism, economism and sociologism. And the common denominator to these reductionist approaches is that ontology (or at least social ontology) reduces to either physics, biology, psychology, economics or sociology. But, just as NE does not conform to physicalism, it does not reduce to any one of the other sciences mentioned. Most certainly NE would be in conflict with the types of reductionism mentioned, except maybe by a particular interpretation of biology. So what is then going on in Nordic ecophilosophy? It is that there is a tendency to take ecology as an ontological model for environmental philosophy. There is a tendency towards the position that nature, humans and society should be understood in light of ecology. In particular this tendency was part of NE at the very beginning. Sigmund Kvaløy Setereng coined the term 'ecophilosophy' as the 'philosophy of the ecological sciences'. And the concepts 'complexity' and 'complication' were introduced to justify a metaphysical distinction between physicalism and ecology. On the basis of such a distinction Kvaløy Setereng was deriving important ecophilosophical implications about nature, humans and society; who we are and what to do (Kvaløy 1971). In his 1971 paper Kvaløy states: 'More and more attention was payed to the concepts in scientific ecology . . . and the possibility of using these concepts to an analysis of society as being a part of the biosphere', and he continues, 'the concept of "complexity" (as opposed to the concept of "complicated") had its basis in ecology and the understanding of "diversity" introduced by the Odum tradition' (1971: 2). He also underlines the fact that the great attention payed to ecology in this particular sense was the ambition of using these concepts to formulate an eco-philosophical approach with relevance to ecopolitics (1971: 3). Later on in the same paper Kvaløy is applying directly the scientific concepts of 'succession' and 'climax state' to society and he says it in the following manner; 'the climax state is an equilibrium state "in which no population or species is able to expand at the cost of others"'. He then concludes; 'this is like the *steady state society* i.e. the society with zero growth in an ecological sense' (1971: 35–6) and which Kvaløy thinks it is worthwhile striving for. The founding father of 'deep ecology', Arne Næss, has a similar reasoning. In his 1989 book he states that in nature 'everything connects with everything else' or the 'relational total field image'. In Næss' terms, ecology 'suggested, inspired and fortified' the metaphysical conclusion that individuals are constituted by their relations to a wider living whole (Næss 1989). From this human responsibility towards any living and non-living system flows, i.e. the ethics of 'inherent value'. With respect to this particular interpretation of Næss Des Jardins says:

> Like the land ethics, Deep Ecology relies on the science of ecology for a variety of purposes. Ecology provides a good deal of information about how natural ecosystems function. Ecology helps us diagnose environmental disorders and prescribe politics that can resolve these disorders . . . Accordingly, ecology contributes to ecophilosophy in the same way as scientific understanding has often contributed to ethical analysis.

Des Jardins' conclusion is; 'Næss believes that ecology can provide a model for thinking about the deep philosophical questions that need to be addressed' (1992: 215–16).

My reason for saying that NE could be taken to be 'reductionist' in this particular sense is that I think it is problematic. First, because no one actually knows what metaphysical or ontological views of nature, humans and society can be derived from ecology (or biology) and second that such an 'ecologism' might in fact be too simple minded when it comes to understanding man and nature and the relation between the two. So what is the problem then?

The problem of basing ecophilosophy in ecology or biology

I have no quarrel with NE as a criticism of the traditional Cartesian or positivist picture of the world or the 'human exceptionalist thesis'. From an ecophilosophical point of view this is important criticism leading our ideas towards new territory as we are facing serious problems or the ecological crisis. However, what my concern in this particular connection is, is to sort out some differences between critical realist philosophy and NE projects. So this is what I am going to focus on in what follows.

One of the problems with ecologism is that there is very little agreement to the fact what scientific ecology actually amounts to. If we look to the history of ecology we can at least talk about three different and partly incommensurable models even if it is a very young science. First we have the *organic model* of Clements which has its very roots in Aristotelian teleology. Second there is the *mechanist model* of Odum and Hutchinson which is inspired by a physicalist approach (or system theoretical approaches) seeing ecosystems as mechanical systems. Third, and just recently developed, is the ecology of *'disturbance and patch dynamics'* which has changed the picture completely with respect to how to understand phenomena at this level in nature (Worster 1993). What is more, all three models are still at work in modern ecological understanding. These three different models can be divided into two main paradigmatic perspectives on nature. Clements and Odums share a common assumption. Natural ecosystems tend towards a point of relative stability or equilibrium, i.e. 'homeostasis' or the 'climax state'. When the system is disturbed, natural forces work to return to a point of equilibrium. A system in equilibrium or 'climax state' (homeostasis) tends to resist change and stay in balance if it is not disturbed and if it is disturbed it will restore its balance if left in peace. The disturbance paradigm on the other hand says that there is rarely any equilibrium state in nature (no homeostasis or climax state). In this latter view natural systems or ecosystems are viewed more or less as chaotic systems. A view which suggests that ecosystems are in constant change and perhaps more important that this change occurs without direction or any sense of 'development' (or succession), as has been a central claim in the traditional view. Nature is just processes and change. Næss and the early Kvaløy Setereng relied very much on the Clements/Odum ecosystem view of life, i.e. the 'balance paradigm'. This should, however, come as no surprise since using the 'equilibrium model' as a guide, environmentalists could in fact reason from the objective facts of ecology to human political and ethical conclusions. Just as we can speak of the health and well-being of an individual organism or a physical system, we can speak of the health and well-being of ecosystem. Or as stated by Næss:

> The ecosystems in which we intervene are generally in a particular state of balance which there are grounds to assume to be of more service to mankind than states of disturbance and their resultant unpredictable and far-reaching changes. In general, it is not possible to regain the original state after the intervention has wrought serious, undesired consequences.
>
> *(1989: 26–7)*

However, the later Kvaløy Setereng seems to have turned away from the 'equilibrium' or 'climax state' perspective towards a more extreme process view of life. Everything is seen as process in time, and everything is changing with time, individuals, society and nature (Kvaløy Setereng 2002). Earlier ecophilosphical concepts such as ecological equilibrium, climax state, ecological structures and steady-state society are thus given up. Hence, the later Kvaløy Setereng sees things more in accordance with the paradigm of 'disturbance and patch dynamics'. His latter views accept that there is relentless diversification, openness and change. We might in fact state that there are at least two different NEs. One in Næss where the overall picture is based on a holistic view of life, where unity and identity with nature and the 'relational total field image' is of utmost importance. The other represented by Kvaløy Setereng where the attention is drawn in favour of understanding nature as diversification, process, openness and change.

However, the process view of life seems to be just as problematic as the 'climax state' or homeostatic picture. That is, if everything is understood as process and change in open systems then there is the danger that nature at this level vanishes altogether. For instance Marc Sagoff and Kristin Shraader Frechette have shown that if we accept the paradigm of 'disturbance and patch dynamics' then scientific ecology cannot tell us anything important about ecosystems (Sagoff 1997). If ecosystems resemble chaotic system then they are not in any equilibrium states (or homeostasis) and if this is so then we cannot talk about conserving or 'sustaining' such systems. The reason is that they are too complex; so that if we intervene we might in fact make things worse. This is because emphasizing the complex interconnections within ecosystems, this view holds that small random changes can have major and unpredictable catastrophic consequences (Des Jardins 1992). So if everything is changing at this level in nature it does not seem to be anything to care about or to be responsible for. Hence, we cannot just refer to scientific ecology to justify our claims about ecophilosophical political or moral concern.

We could of course use biology or biological systems in general as our ecophilosophical grounding and then argue for the view that we should take those necessary measures so as to preserve the natural world including ecologically relevant objects and systems. But it seems that such a point of view is even more problematic. The reason being that, even if there is a sense in which we can say that Darwin's theory of evolution by natural selection is accepted as the only qualified interpretation of the history of life on our planet, there is nevertheless not much agreement about what follows from such a view; neither about what nature is, what humans are or what society is. For instance, there is the Mayr functionalist interpretation which tells us that the theory of evolution by natural selection does not give us any justification for reductionism, essentialism, determinism or mechanism in any traditional sense (Mayr 1991). But Mayr's approach is just one among several. Dawkins, for instance, argues that the only important thing in the history of life is genes, i.e. it is the 'gene eye view' of evolution (Dawkins 1976). But Dawkins' gene reductionist thesis has been heavily criticized by Stephen J. Gould and Richard Lewontin. Gould and Lewontin pay attention to the fact that evolution is a more complex process that involves many factors and hence cannot just be reduced to 'gene propagation' (Gould and Lewontin 1979). As in Stuart Kaufmann's complexity science, Gould and Lewontin pay attention to holism, interconnectedness and self-organization as forces which are equally important and operative in the evolutionary process.

To see what a mess we are dealing with at this level of nature, Kate Soper is a good source:

> to attend to the symbolic dimensions is immediately to be struck by the diversity and complexity of our descriptions and images of nature. Nature is a machine and an organism, passive matter and vitalist agency. It is represented as both savage and noble, polluted and wholesome, lewd and innocent, carnal and pure, chaotic and ordered . . . Sublime and pastoral, indifferent to human purposes and willing servant of them, nature awes as she consoles,

> strikes the terror as she pacifies, presents herself as both the best of friends and the worst of foes.
>
> *(1995: 71)*

Taking biology/ecology as our favourite science seems too hasty. There is no agreement as to the metaphysical or ontological implications or commitments which can be drawn from these sciences. Hence, it also seems problematic to found ecophilosophy or ontology on these sciences. We need to develop alternative approaches which enable us to transcend biologism and ecologism.

The 'epistemic fallacy'

What are the perils of biologism and/or ecologism in relation to ontology. To see the problem we need to introduce Bhaskar's notions of the 'epistemic fallacy' as well as the 'ontic fallacy'. That is to say there is a tendency in NE to derive ontological claims from their basis in knowledge, or 'the analysis of statements of being in terms of statements about our knowledge of being'. It effects an *epistemologization* of being, i.e. something that 'commits to the category mistake of conflating things and thought' (Hartwig 2007: 173).

But there is also the vice versa, i.e. since ontology is in fact irreducible to epistemology, the epistemic fallacy merely masks the generation of an implicit ontology; 'establishing a drive to subject-object identity theory', where it entails the 'converse ontic fallacy, viz. the definition, or assumption of the compulsive determination, of knowledge by being, or the *ontification* or the *ontologisation* of (current) knowledge' (Hartwig 2007: 174).

Or as said by Bhaskar:

> philosophical ontologies . . . must be distinguished from specific *scientific ontologies* constituted by particular scientific theories, postulating particular types of entities and processes, which detail the specific features and particular types of the general landscape which philosophical ontology demarcate and charts.
>
> *(2002b: 26)*

To provide biological or ecological 'facts' as evidence for ontological claims does not seem to be quite appropriate.

To see the ontological flaws of this specific type of reductionism or the tendency towards ecologism I want to introduce an alternative critical realist perspective. First I want to show that basic critical realism enables a critique of all kinds of reductionism based on its arguments of an ontology which argue for a stratified and layered reality with emergent powers in open system, i.e. natural necessity, both natural and social. I will then proceed so as to see how dialectical critical realism and the philosophy of metaReality are able to overcome or transcend the two different NEs.

Basic critical realism

Critical realism is a realism about the world. A realist philosophy is committed to the existence of a real world, including natural necessity, natural and social, which exists and manifests itself independently of our knowledge of, or our beliefs about, it. CR holds that this world is in principle knowable, but that our knowledge of it might change as we proceed in scientific investigation. One of the reasons Bhaskar came to such a realist conclusion was in relation to 'asking the important question of what does it consist to make sense of science'; and the answer to that question was 'only on the assumption that the sciences *are about* something which exists independently of their study'. Second, there is in critical realism the attention paid to *reflexivity*, that is, 'about the possibility for thought or ideas to represent something outside itself, science is a social activity and might involve variable means of representation' (Benton and Craib 2001: 120). This is to say that we do not have any absolute certain knowledge of the world as it is independent of knowledge (epistemological relativism) but nevertheless our knowledge is about a world which is 'outside' that knowledge (ontological realism). Most importantly critical realism holds that we must always be open to corrections about our beliefs about the world (judgemental rationality), they can be wrong. CR is thus fallibilist in contrast to positivist or social constructivist theories of knowledge which either rely upon a correspondence theory of truth or insulate themselves from having knowledge of a world which we might be wrong about.

Due to this particular distinction between our knowledge of the world (transitive dimension) and the world as it is independent of that knowledge (intransitive dimension) Bhaskar is able to argue for a 'depth realism', i.e. a natural necessity that we are not able to detect just by scratching the surface of things. This is due to the reality of the intransitive dimension. We need to go behind mere appearances so as to understand what is going on in science. Bhaskar is arguing for transcendental realism, a realism of natural necessity, which goes beyond what we are able to observe at any particular time. To be able to clarify such a 'depth ontology' CR posits three domains of being. There is the *real* world of mechanisms, powers, tendencies, which science seeks to discover. There is the *actual* level of sequences of events that happen in the world caused by the underlying real level and includes both those phenomena observed and not observed by anyone (an avalanche in a remote mountain area not seen by anyone for instance). Then there is the level of the *empirically* observed events which is but a tiny part of what happens at the actual level. What Bhaskar is arguing for then is a stratified and layered reality: 'And the metaphor of levels indicates that critical realism is a form of "depth realism", such that science attempts to discover what is behind the surface appearances of things and uncover their generative causes' (2001: 125).

The idea of reality as layered and stratified can be taken further. To the extent that the different sciences are each concerned with particular levels of reality they too can be ordered into a hierarchy of levels such as that of social science, psychology, biology, chemistry and physics. The important part of Bhaskar's understanding of such a layered or stratified reality is that it represents what he calls an *integrative pluralism*. Such an integrative pluralism allows that each level have its own powers and tendencies that cannot be reduced to the level below. Hence, there are also important reasons why reductionism does not work. First and foremost because 'the lower level sciences explain (at best) only the constitution of the mechanisms at the higher level'. It does not for instance explain when or with what effects the powers established by those mechanisms will be exercised (an example is how we learn language). Second, 'once higher level mechanisms are formed, their activities have effects on the lower level ones', i.e. we have 'holistic causality' (for example emotional trauma might affect respiration). Finally, 'the association between levels and particular sciences is in part explained by the way entities at higher levels have properties and powers not predictable in advance on the basis of properties of lower level entities' (we need to know about higher level entities (genes for instance) before we can try to understand what they are composed of (constellations of molecules)) (Benton and Craib 2001: 126).

The particular important debate around reductionism as we have mentioned is concerned with the higher level sciences, i.e. whether sociology can be reduced to psychology and psychology eventually reduced to human biology, etc. In relation to such a reduction of the human sciences to lower level sciences Bhaskar develops what he calls a *synchronic emergent powers materialism* (SEPM) (Bhaskar 1986: 107) which 'suggests that the objects of the (human) sciences are distinct (and perhaps emergent) from, but connected to, and in particular grounded in, nature'. What he is stating here is that they are 'taxonomically and causally irreducible, but dependent, modes of matter' and there is the following characteristics between the layers: first there is *Unilateral Dependency* – the objects of human sciences are unilaterally existentially dependent on those of the natural sciences, secondly; there is *Taxonomical Irreducibility* – the natural sciences are at present unable to explain the human world under human descriptions, and thirdly there is *Causal Irreducibility* – reference to properties not designated by physical theory is (apparently) necessary to explain some physical states (viz those resulting from 'intentional action' (ibid., p. 107)). Bhaskar, however, underlines the fact that SEPM does not require the postulation of any

> substance other than matter as bearer of the emergent powers, so that it is not dualist; and it is consistent with a diachronic explanatory reduction, i.e. a natural-historical explanation of the formation in time of emergent powers, so that it is not preformationist or creationist, i.e. it is metaphysically Darwinian.
>
> *(1986: 107)*

Due to the depth reality and the ontology of natural necessity in a stratified and layered reality within emergent open systems, reductionism cannot and does not function. Hence, to use reductionist methods to study phenomena must be highly inadequate or inappropriate. Thus, a 'tout court' reductionism of most kinds is seriously flawed.

Dialectical critical realism

Having done away with reductionist approaches we are now able to go into the question as to how critical realism is able to transcend and unify the two pictures picked out by NE, i.e. the holistic, identity and unity in difference, and the 'total relational field' view developed by Arne Næss and the process oriented, diversity, open change view developed by Kvaløy Setereng. As I see it, in dialectical critical realism and in reflections on metaReality Bhaskar is able to fuse these two positions into one overarching view which replaces and further develops the two NE positions. To see how this is possible I will here make a brief outline of some central concepts developed in dialectical critical realism.

As stated by Alan Norrie in his book on dialectical critical realism; 'Critical realism's central commitment, emerging out of asking the question about how knowledge is possible, is to the idea of an ontology endorsing depth and natural necessity – to a theory, in short, of the ontological necessity of being.' Norrie then asks; 'If that in brief is critical realism, what does its dialectical form add' and his answer is 'It will integrate an account of becoming with the existing critical realist understanding of being' (Norrie 2010: 13).

According to Norrie, then, the key development in dialectical critical realism

> is to take the theory of a structured nature of being (natural necessity) and to think about how being changes in time and space, and how philosophically that change is to be understood . . . The core move of Bhaskar's *dialecticisation* of critical realism is thus the development of his realist account of ontology as natural necessity in terms of the relationship between being and not-being, and therefore being and becoming.
>
> *(2010: 15)*

This particular understanding of a reality in change and development from being to becoming is to be understood in relation to the first two stages in the MELD schema (Bhaskar and Hartwig 2010: 123–4). In this regard Bhaskar's dialectics starts with the 'first moment' (1M), 'with the sheer, real difference that exists in the world (what he calls non-identity)'; and this is:

> Because all the basic distinctions within transcendental realism turn on non-identity relations. The object is not the same as the knowledge of the object, this is the transitive/intransitive distinction, then you have the distinction between the domains of the real, actual and empirical. All these involve non-identity relations, and that supplies the logical infrastructure of basic critical realism.
>
> *(Ibid.: 123)*

He then moves to the 'second edge' which opens up for further deeper levels of ontology, (2E) is the realm that is opened up by absence which includes real negativity and contradictions that things in the world disclose. But there is also the level of totality, internal relationality and holistic causality; this is to locate negativity in the world at the 'third level' (3L) of a world of process, negation and transformation of a real, open unfinished whole (totality). 'So while at 1M (you) think being as non-identity and structure, at 2E you think it as process. Then at a third, deeper level, 3L you think being as whole, so critiquing ontological extensionalism by seeing internal relations between the different parts of being, not just external connections' (ibid.: 124). Bhaskar then argues for the importance of a 'fourth dimension' (4D) of transformative praxis, i.e. the capacity for practical human agency to change the world.

The above described movement of being to becoming thus deals with ontology and being as developed in original critical realism and of absence and becoming as developed in dialectical critical realism. This concerns for the most part being as objects and structures which through concepts such as absence and negativity are in continuous dialectical change individually as well as in relation to other things and objects. But a realist depth ontology (natural necessity) is also related to 3L of totality.

> Natural necessity thus leads to a conception of structured totality and, in the light of the workings of real negativity (change, process, becoming) in the whole, presents a conception of a totality that is in motion. The critical realist totality is thus also a process of becoming, is incomplete and assumes an open future.

And as further stated by Norrie, 'This must be understood as a realist, emergent sense of open totality' (2010: 16).

The fourth dimension, 4D, in dialectic sees the importance of agency as transformative practice. One core idea in this dialectic is that 'desire grounds the existential experience of human being as at once governed by natural necessity (natural and social), and a being in search of freedom'. As Norrie further claims:

> When pushed to their limits, the conflicting demands of this experience push the individual to an idea of human freedom and emancipation as the utmost possible satisfaction of desire, and therefore, in so far as it is possible for material beings, liberation from necessity. This is given a universal quality by the realization that the freedom of each depend on the freedom of all . . . What we have is a process of dialectical rationality in which thinking consistently from one's concrete reality as a needy being to the ultimate satisfaction of one's ends entrains the thought of a eudaimonic world for all.
>
> *(2010: 122)*

In 4D Bhaskar thus

> develops a complex scheme of ethical elements involving what it means to be free, to trust, and to enter into solidarity with others. He talks about

> principles of dialectical rationality which press human beings, through the necessary universalisation of their practical commitments, towards ethical positions. These lead them eventually to the idea of an eudaimonic society, the best of moral worlds in which a fully developed understanding of the freedom of each is achieved only though the freedom of all.
>
> *(Norrie 2010: 17)*

An achievement of such an eudaimonic world is dependent on our human agency and the struggles in which we must partake. In this regard, individuals must be seen as part of an open future in a structured plurality and totality. The future cannot be closed. It is open, in process of development, and subject to substantial change and the qualitatively new. Something that 'proceeds from our desire to be free from material needs to an eventual desire by extension for freedom of a most complete form' (Norrie 2010: 122).

metaReality

In his 'Introduction to Reflections on metaReality' Hartwig says (2011: 2) that Bhaskar's philosophy has been elaborated in three main phases, each developing and deepening its predecessor: original or basic critical realism (CR); then 'dialectical critical realism (DCR) (the dialecticization of critical realism and the emancipation of dialectic (for the dialectic of) emancipation; finally; the philosophy of meta-Reality'. Here we need to understand that the philosophy of metaReality (MR) is

> preservative sublation of CR/DCR: it both draws out its real strengths and, without falling back into identity-thinking, goes beyond it by reversing its prioritization of difference (non-identity) over unity and identity; in which the earlier system remains valid as an account of the fundamental shape of relative reality (the world of non-identity and duality) but is surpassed as 'realism about transcendence leads to self-transcendence of realism' in a conception of an infrastructural absolute reality or foundational level of being that, as a necessary condition 'for any being at all'.

According to Hartwig, metaReality represents Bhaskar's spiritual turn:

> it was motivated above all by the desire to find a way forward out of the multiple interrelated crisis affecting human social being at all four of its plane (our transaction with nature, interpersonal relations, social structures and stratified personalities all of which according to Bhaskar is out of kilter with our fundamental natures).
>
> *(2011: 3; see also Bhaskar 2002a: 22)*

Thus metaReality involves the further deepening of ontology, where primacy is to 'identity over difference and unity over antagonism and split' (Bhaskar

2002a: 18). As stated in 'Formation . . .'; 'I saw that it was necessary to have further levels' (referring to the four levels in the MELD scheme he adds three more levels 5A, 6R and 7A/Z) and he further states that: 'at 5A you think being as spiritual or as incorporating self-reflexive human praxis; at 6 R you think it as (re-) enchanted or as meaningful and valuable in itself; and then at 7A/Z you think being as non duality, more generally as one' (Bhaskar and Hartwig 2010: 123). And the development is a necessary one:

> The first four steps (MELD) are the crucial ones for understanding what I have later called the world of duality, including specifically the world of demi-reality. They are perfectly adequate for understanding that world, but not necessarily for bringing about the change we want, which require the further deepening of ontology in the last three steps.
>
> *(Bhaskar and Hartwig 2010: 124)*

In this regard metaReality gives us access to our deeper selves or our 'ground states' and how these ground states contain the possibility of emancipation and freedom. Bhaskar says:

> Now beneath that embodied personality (and this is my argument) there is the essential, transcendentally real self, which is not separate from other selves or other things; which is in fact continuous with and connected to all other selves and other things generally and indeed to all of creation. This is the transcendentally real self, and this is the self, the 'I' of causal agency. It's at this level things happen in the world. So this deep essential self, which I've also called our alethic self, that is our true real self. This real thing is connected with everything else in the world. It is the source of our energy, creativity and intelligence; our love, joy and freedom; our spontaniety, unconditionality and agency. Nevertheless it is concrete and singularised. It is a particular; and just as we have unique personalities, everyone is different in their innermost being.
>
> *(Bhaskar 2002a: 78)*

Hence:

> The task of emancipation is to shed: to shed what we are not and to fully embody everything that we essentially are . . . to become our transcendentally real self, to actualise that human creativity, that source of energy and intelligence, capacity for love and right-action which sustains everything.
>
> *(2002c: 86)*

As said by Cheryl Frank, Bhaskar wants us to be in touch with our ground states or our best natures, or higher selves, which no one can lose because it is part of being human, no matter how conflicted or split a person becomes. And as people get in touch with their best natures, they also want to act to ensure the earth, and the life systems it supports, endures (Frank 2010: 113).

In relation to our ecophilosophical concerns then getting in touch with our 'best selves' is then to actualize 'the capacity for love and right action which sustains everything', nature included. The 'main point' is well grasped in a quote from Albert Enstein:

> A human being is part of a whole, called by us the universe, a part limited in time and space. He experiences himself, his thoughts and feelings, as something separated from the rest, a kind of optical delusion of his consciousness. This delusion is a kind of prision for us, restricting us to our personal desires and to affection for a few persons nearest us. Our task must be to free ourselves from this prison by widening our circles of compassion to embrace all living creatures and the whole of nature in its beauty. The true value of a human being is determined by the measure and the sense in which they have obtained liberation from the self. We shall require a substantially new manner of thinking if humanity is to survive.
>
> *(Hartwig 2011: 11)*

Concluding remark on NE and critical realism

Frank summarizes Bhaskar's contribution nicely in the following statement:

> Bhaskar discusses and analyses the many levels or dimensions of ontology: being as such and being as structure, being as process and change that incorporate absence and negativity and contradictions, being seen as involving totalities, being as transformative human agency, being as reflexive and spiritual, being as transcendence or 're-enchantment' of a world that is meaningful and valuable in itself, and all being as part of a wholeness or unity in diversity, with depth and variety which continues to enfold into new aspects.
>
> *(Frank 2010: 103)*

The problems inherent in the two different NE ecophilosophical accounts might lie in the fact that they are too close to biology and ecology. First and foremost such a view is problematic not the least due to the inability to differentiate knowledge from being, by defining being in what we can possibly know, i.e. NE has not elaborated on in depth analysis of the distinction between the transitive and the intransitive dimension. Second, by not being able to differentiate between being as process and change and being as part of a wholeness or unity in diversity NE gets to be too simple minded when it comes to understand who we are and what to do.

In its attention to ontology and the interrelated network of conceptual clarification, Bhaskar's perspective is not just a synthesis of the two NE perspectives. Critical realism gives us a more comprehensive account of the real world through its development of a philosophical depth ontology. By then further developing this ontology in dialectical critical realism through the extended MELD system, and subsequently deepening it in the philosophy of metaReality, Bhaskar is able to transcend and develop further the correct but limited intuitions of Nordic ecophilosophy (NE).

Anthropocentrism vs. non-anthropocentrism

In conclusion I want to extend my arguments not only to cover NE projects but also to have consequences more generally for anthropocentric (as distinct from non-anthropocentric) approaches. I take both anthropocentrism and non-anthropocentrism to rely upon different types of reductionist naturalism. In this regard ontological actualism extends to ethical reductionism. Taking into consideration different anthropocentric as well as non-anthropocentric points of view such approaches derive from scientific ontology (biological as well as ecological) ethical claims about what has moral standing or what might have 'inherent values' or values in themselves. First most anthropocentric approaches are reductionist in the sense of accepting the radical dualism between man and nature on the basis of physicalism. This is the classical view put forward by such philosophers as Descartes and Kant and which is followed up by modern anthropocentric approaches (Baxter). Our moral concern cannot be extended to include more than the human world because if nature is nothing more nor less than physicalist matter or our own 'construct' there is nothing 'out there' to be morally concerned about. Hence nature has only instrumental value, values relative to human concerns. Reductionist naturalism can be argued for also in relation to different non-anthropocentric approaches. Singer's and Regan's animal rights approach reduces moral concern to exclude anything but animals, i.e. what is 'sentient' or 'subject of a life'. The same type of reductionism is involved in biocentrism as developed by Taylor, Hans Jonas and others where nature's 'flourishing' becomes the meta-criterion. Leopold reduce ontology and ethics not to individual things, humans or natural, but to whole systems such as landscapes, mountains, rivers and ecosystems. Leopold's ethics were motivated from and inspired by the theory of evolution. However, the common denominator in anthropocentric as well as in non-anthropocentric approaches is that ethics develops as a result of definite criteria, which have their basis in particular interpretations of physics (matter in motion), biology (animals or life systems in general) or in ecology (ecosystems and landscapes). This is ethical reductionism based on different actualist interpretations of scientific ontologies. If things do not conform to these definite criteria they are not a part of the moral community. However, these particular interpretations are themselves, to use Bhaskar's language, committed to the epistemic fallacy, i.e. deriving ontological conclusions from scientific 'facts'. The epistemic fallacy here is part of what Bhaskar calls the 'anthropic fallacy', i.e. if the epistemic fallacy misunderstands beings in terms of human knowledge, the anthropic fallacy misunderstands it in terms of human beings as such. That is to say that traditional anthropocentric as well as non-anthropocentric approaches conflate ontology with epistemology on the basis of anthropocentrism, i.e. the view that humans are the centre around which everything else revolves and hence are the creatures which decide what is to fall under the term moral standing or not. This is the case whether we are talking about animal rights, living systems rights in general or rivers, landscapes and ecosystem rights. Humans becomes the origin of what counts as valuable 'in itself' or not, anthropocentric or non-anthropocentric. Non-anthropocentric approaches merely mask anthropocentrism by virtue of extending what can have moral standing to the non-human world, as it still is us who are doing the carving, i.e. it is we who are defining what is to count as values in themselves even if we extend this to include animals, plants and ecosystems. Hence different non-anthropocentric approaches are still anthropocentric in the sense that humans are the ultimate arbiter when it comes to what should be worth our moral concerns or not.

In opposition to this, Bhaskar is a 'full blown realist'. He is 'carrying through a real "Copernican revolution" that decentres humanity from the universe and underpins a comprehensive new realist philosophy' (MinGyu 2008: 9). Bhaskar is 'reclaiming reality of independently existing things and transfactually efficacious real structures, systems, mechanisms, processes, wholes, relations and fields which exist independently of human beings' in the broadest possible sense. This in turn requires the development of a 'new understanding of the relation between human beings and reality (nature). Reality (nature) itself is the structures and emergent process of the universe (natural necessity) and humans are one of the elements of or participants in the process, not the organizer or the creator of it' (2008: 10). MinGyu states:

> Bhaskar here is more in accord with East Asian philosophy because unlike the secular humanism of western thought the Confuscian *anthropocosmic* way is based on the view that human beings are not the centre or the unique subject in nature but just a participant in or part of nature. In other words, human being and human creativity are subsets of the self-realisable possibilities of the enfolding cosmos or heaven. In this regard the Eastern concept of nature has as its centre the idea of an all-embracing self-generating process of life including human being.
>
> *(2008: 10)*

In this regard, 'natures forms pertains the most fundamental level of the structured intransitive' (natural necessity) (ibid.: 10), and human beings are essentially 'ground states, our transcendentally real selves located at the cosmic envelope with potentially cosmically infinite powers, inseparably connected with all other beings in the cosmos' (ibid.: 15). In short, anthropocosmic human beings and nature are meta-Reality ground state (the transcendentally real self) and cosmic envelope (the unity of ground state) respectively. So only in Bhaskar's metaReality system is the possibility of 'incorporating the whole of nature preserved as potentiality; indeed the whole of nature always already is incorporated in human nature (and the nature of other beings) as implicit or enfolded potential' (2008: 14). Hence, it is in this sense we must understand Bhaskar's anti-anthropocentric or anthropocosmic view of 'being as transcendence or "re-enchantment" of a world that is meaningful and valuable in itself' (see 6R).

Thus, from critical realist ontology there is the possibility to 'revisit the question of moral community and moral agency that has caused concern in environmental ethics' (Parker 2010: 210). What is required for such a revisit is exactly not to resort to anthropocentrism in our view of the world and our place in it. By accepting non-reductionist as well as non-biologist and non-ecologist approaches and open up for a more advanced ontology (natural necessity, the structured intransitive) and a more advanced non-anthropocentric approach we can then see how critical realism might improve upon both traditional anthropocentrism and non-anthropocentrism as well as Nordic ecophilosphical projects by providing us with a more profound, diversified and richer as well as unifying ontology and a more profound ethical approach that is able to lay the ground for better decision-making and which might 'include the greater prospects of an integrative self-understanding that could be truly emancipatory' (Parker 2010: 210).

Bibliography

Barry, J. *Environment and Social Theory*, Routledge, London, 1999.

Benton, T. *Natural Relations*, Verso, London, 1993.

Benton, T. and Craib, I. *Philosophy of the Social Science*, Palgrave, New York, 2001.

Bhaskar, R. *A Realist Theory of Science,* Harvester Wheatsheaf, London, 1978.

Bhaskar, R. *The Possibility of Naturalism,* Harvester Wheatsheaf, London, 1979.

Bhaskar, R. *Scientific Realism and Human Emancipation*, Verso, London, 1986.

Bhaskar, R. *Reclaiming Reality*, Verso, London, 1989.

Bhaskar, R. *From Science to Emancipation*, Sage Publications, New Delhi, 2002a.

Bhaskar, R. *From East to West*, Routledge, London, 2002b.

Bhaskar, R. *Reflections on Meta-Reality – Transcendence, Emancipation and Everyday Life*, Sage Publications, New Delhi, 2002c.

Bhaskar, R. and Hartwig, M. *The Formation of Critical Realism: A Personal Perspective*, Routledge, London, 2010.

Dawkins, R. *The Selfish Gene*, Oxford University Press, Oxford, 1976.

Des Jardins, J. *Environmental Ethics*, Wadsworth Pub. Company, Belmont, CA 1992.

Frank, C. 'Global Warming and Cultural/Media Articulations of Emerging and Contending Social Imaginaries; a Critical Realist Perspective', in *Inderdisciplinarity and Climate Change* (ed. Bhaskar *et al.*) Routledge, London, 2010.

Gould, S.J. and Lewontin, R. 'The Spandrels of San Marco and the Panglossian Paradigm, a Critque of the Adaptionist Program', in *Proceedings in the Royal Society of London*, 1979.

Hartwig, M. *A Dictionary of Critical Realism*, Routledge, London, 2007.

Hartwig, M. 'Introduction to Reflections on metaReality' (forthcoming).

Jakobsen, T.G. 'Funksjonell Realisme og Dawkins Gener', *Norsk Filosofisk Tidskrift nr. 1–2* 2003.

Jakobsen, T.G. 'Realism, Konstruktivisme og Pluralistisk Realisme', *Naturen nr. 2,* Universitetsforlaget, Oslo, 2004.

Jakobsen, T.G. *Økofilosofi – økologi, evolusjonsteori og transformativ læring*, Tapir, Trondheim, 2005.

Kvaløy Setereng, S. *Økofilosofisk Fragment*, Oslo, 1971.

Kvaløy Setereng, S. *Mangfold og Tid*, Tapir, Trondheim, 2002.

Leopold, A. *A Sandy County Almanac*, Oxford University Press, Oxford, 1949.

Mayr, E. *One Longue Argument*, Harvard University Press, Cambridge, MA, 1991.

Mayr, E. *What is Biology?*, Harvard University Press, Cambridge, MA, 2001.

MinGyu, Seo, 'Bhaskar's Philosophy as Anti-Anthropism: A Comparative Study of Eastern and Western Thought', *Journal of Critical Realism*, Vol. 7, No. 1, 2008, (pp. 5–28)

Næss, A. *Ecology, Community and Lifestyle*, Cambridge University Press, Cambridge, 1989 (*Økologi, Samfunn og Livsstil*, Oslo, 1974).

Norrie, A. *Dialectics and Difference*, Routledge, London, 2010.

Parker, J. 'Towards a Dialectics of Knowledge and Care in the Global System', in *Interdisciplinarity and Climate Change* (ed. Bhaskar *et al.*), Routledge, London, 2010.

Plumwood, V. *Feminism and the Mastery of Nature*, Routledge, London, 1993.

Plumwood, V. *Environmental Culture, the Ecological Crisis of Reason*, Routledge, London, 2002.

Regan, T. *The Case for Animal Rights*, University of California Press, 1983.

Sagoff, M. *The Economy of the Earth*, Cambridge University Press, Cambridge, 1988.

Sagoff, M. 'Impediments to a Theoretical Ecology', *American Phil. Assoc.*, 1997.

Session, G. (ed.) *Deep Ecology in the 21st Century*, Shambala, Boston, 1995.

Singer, P. *Animal Liberation*, Random House, New York, 1990.

Soper, K. *What is Nature?*, Blackwell, Oxford, 1995.

Sutton, P.W. *The Environment, A Sociological Introduction*, Polity, Oxford, 2007.

VanDeVeer, D. *People, Penguins and Plastic Tree*, Wadsworth, Belmont, CA, 1993.

Warren, K. *Ecofeminist Philosophy, A Western Perspective on What It Is and Why It Matters*, Rowman and Littlefield, Oxford, 2000.

Wilson, E.O. *Conscillience*, Harvard University Press, Cambridge, MA, 1998.

Wilson, S. and Sober, E. *Unto Others*, Harvard University Press, Cambridge, MA, 1998.

Worster, D. *The Wealth of Nature*, Oxford University Press, Oxford, 1993.

14

FROM ECOPHILOSOPHY TO DEGROWTH

Karl Georg Høyer and Petter Næss

This chapter traces the contributions to debates on economic growth and the ecological crisis from the first generation of Nordic ecophilosophers (in particular Arne Næss and Sigmund Kvaløy and their predecessor Peter Wessel Zapffe), via significant Finnish (G.H. von Wright) and Swedish contributors (T. Hägerstrand), up to contemporary writers (notably Arne Johan Vetlesen). More or less explicitly, the Nordic ecophilosophers were all generally critical to growth, although some of them seemed to be more worried about population growth than economic growth per capita. Kvaløy considered 'Industrial Growth Society' as the antidote to his ideal of a 'life necessity society'. Similarly, Næss held economic growth as conceived and implemented by industrial states to be incompatible with his deep ecological principles, although he did not rule out the possibility of a fundamentally different, ideal form of sustainable economic growth. The Nordic ecophilosophers also differed somewhat in their view on whether or not economic growth is indispensable for a capitalist economic system. In this chapter, we will focus on the ontological and ethical bases for the growth criticism of the Nordic ecophilosophers. Links between the Nordic ecophilosophical tradition and the emerging European-scale degrowth movement will also be explored.

The historical foundation of the economic growth critique

During the last decade, in particular, an international discourse on *degrowth* has strongly evolved. With this term as the sole focus a first international research conference was arranged in Paris in 2008, and a second one in Barcelona in 2010. The degrowth discourse is much in line with the basic economic growth critique highlighted by the Nordic ecophilosophers.

The American-Rumanian theoretical economist Nicholas Georgescu-Roegen (1971, 1976) is considered by many 'the father of degrowth' (Martínez-Alier *et al.*,

2010). Degrowth, as the English translation of La décroissance, is associated with the notion of 'declining' as used by Georgescu-Roegen (1976) in his book collection of papers, *Energy and Economic Myths*. In this he debated the report on *Limits to Growth* by Meadows *et al.* (1972) and the thesis about the steady-state economy championed by Herman Daly (1972). It is fair to say that Georgescu-Roegen was the first economist to clearly recognize the relevance of thermodynamic constraints, especially the entropy law, to economics. According to this second law of thermodynamics energy in isolated systems transforms towards ever higher degrees of disorder, or entropy which is the term for disorder; thereby the entropy law. Georgescu-Roegen's thesis about the necessity to understand thermodynamics in order to think adequately on economics was endorsed early and discussed by important French authors like André Gorz (1975). The French translation and edition by Grinevald and Rens (1979) of a selection of Georgescu-Roegen's writings with the title *Demain la décroissance* (*Degrowth for Tomorrow*), a book reprinted in more extensive versions in 1995 and 2006, had very important impacts in the academic and ecological movement circles in France. This was to be one of the main keys to the establishment of the whole 'Degrowth movement', which is, however, still mainly based in Southern Europe, France, Italy and Spain.

However, long before that, the Georgescu-Roegen writings were a great source of inspiration for the early development of Nordic ecophilosophy, illustrating his key role in the basic growth discourse through more than forty years. Actually, Georgescu-Roegen could in many ways be considered an ecophilosopher, not the least through his keen interest and knowledge in basic ontological and epistemological issues and his deeply rooted interdisciplinarity, integrating theoretical economics with knowledge from physics and ecology. At an early stage he coined the term 'bioeconomics', which is quite similar to our later term 'ecological economics'. Thus extensive references to his writings are found in publications by Arne Næss (1989), but more consistently over time by Sigmund Kvaløy (1974). However, already in his doctoral thesis from 1941 and through several contributions during the 1940s and 1950s, Peter Wessel Zapffe with his basic tragic view on the human species expressed similar insights as Georgescu-Roegen into the thermodynamic entropy law. Actually, the basic principles of thermodynamics have been lay knowledge in human societies for a very long time indeed. Practical applications are known from many ancient cultures thousands of years ago.

Thermodynamics as a science and as laws of physics were firstly identified and described during the nineteenth century. It came out of the interest for steam engines, and the need to fully understand why and how they worked in order to increase their efficiencies. The great discoverer was the French engineer Sadi Carnot (1796–1832). From him we have inherited terms such as the *Carnot-cycle* for heat engines and *Carnot-efficiency*. Through his analyses of the works by Carnot the German physicist Rudolf Clausius (1822–1888) coined the term *entropy* and identified the entropy law, or the *second law of thermodynamics*. He noted that reversibility might still be true for such idealized processes as the Carnot-cycle, but that any real processes had to be regarded as irreversible. The second law thus expresses this basic irreversibility. When expressed it aroused a lot of controversies. To some extent, these controversies have been there ever since. The irreversibility claim was countering the hegemonic Newtonian mechanics and the very basics of the mechanistic paradigm in science. Similarly it is countering the later development of quantum and statistical mechanics. This is the background for one of the key critiques by Georgescu-Roegen of neoclassical economics, which he claims to solely be a mechanistic discipline founded on an illusion of reversibility in real life.

The first law was actually expressed after the second one, and a zeroth law even after that, as a basic necessity for the two other laws to apply. We shall here concentrate on the first and second laws. In this type of physics there is a triple bottom line of systems: *isolated*, *closed* and *open*. This is a somewhat richer typology than the distinction in basic critical realism between closed and open systems. In an isolated system neither matter nor energy is entering or leaving the system boundaries. The closed system refers to one that has energy but no matter crossing its boundaries, while both energy and matter are crossing in the case of open systems. At the outset laws of thermodynamics apply to isolated systems, while the later discourse – not the least as initiated by Georgescu-Roegen – mostly has been about their more extensive and absolute validity in closed and open systems. Also in opposition to classical mechanics thermodynamics highlights *transformation* and *change*. There is thus more resemblance to Aristotelian physics claiming that movement and change was one and the same thing, just the physics Galileo in particular put a lot of ideological efforts into combating. In the mechanics of Galileo and Newton movement and change are two completely different things, and movements can go on with complete reversibility disconnected from any change. According to Georgescu-Roegen, this is also the basic idea in neoclassical economics.

The first law of thermodynamics states that energy never can be created nor destroyed, only transformed from one form to another; or: the energy of an isolated system is constant. When one form of energy is transformed to another form, the total quantity of energy remains the same. Thus the first law is also termed the *law of constant energy*. It has some immediate interesting implications. First, energy-*production* and -*consumption* are not really correct terms; in quantity energy is neither produced nor consumed. The more correct term is *energy use*. However, there are important aspects of energy that are both produced and consumed. This is governed by the second law. A second implication is that all energy use implies transformations in one form or another, and since there cannot be changes in quantity there must be changes in quality. Again, this brings us to the second law. Third, as regards energy at least, something can never become nothing and, conversely, nothing can ever become something. To use the football language: *we can never win*. But we can still hope for a draw, just as claimed in Newtonian physics. Then, at the outset the first law is not really counteracting the basics of mechanism (Høyer, 1979, 1995).

Included in the Georgescu-Roegen theory and the related discourse is the first law extension *to matter*. There is not much controversy about that. As for energy matter can neither be created nor destroyed, only transformed from one form to another. The amount of matter in an isolated system, as well as in a closed system, is constant. The extended law states the *constancy of energy and matter*, and the correct term is *matter use* not production and consumption. When transformed, changes are not in quantity but in quality. There are important relations between the two. Energy requires matter; it is based on matter transformations, just as claimed by Einstein. But matter does not require energy, or does not turn out of energy alone. Energy is however required to make matter *matter* (Høyer, 1995).

Georgescu-Roegen and his fourth law of thermodynamics

According to the second law we cannot even hope for a draw; we are *doomed to lose*. This was the exact insight of the early Norwegian Zapffe's ecophilosophy and the thesis of the human as a tragic species; not only do humans understand that they are doomed to lose, but that they only contribute to ever larger losses through their eternal strive for more industrial affluence. To return to football; there is a vast difference between losing by 0–10 and 0–1, where in the latter case you are all the time at least on the verge of achieving a draw. As emphasized, the second law brings in the concept of *entropy*, thereby the *entropy law*. Entropy is a measure of disorder. Energy with high entropy is in a very disordered form. Conversely: energy with low entropy is in an ordered form. Energy in its ultimate form as dissipated low temperature heat energy has very high entropy; it is unavailable or bound energy which is of no use. It is energy with low or no quality. Energy with low entropy, however, is available energy; it is available both to plants, animals, humans and industrial structures. It is energy with high quality. But any use or application of available energy implies a transformation and loss in quality. Further applications move down the quality ladder ultimately to be transformed into high entropy disordered heat energy. There is no reversibility to this process. Efforts to collect – or concentrate the high entropy energy – will only cause further losses in available energy and more generation of entropy. The second law pervades science, technology and everyday life. It is manifest as friction between moving parts in machines. It sets an upper limit on the efficiency of power plants. 100 per cent efficiency is impossible, and even the maximum Carnot-efficiency of 80 per cent is only possible if the process takes infinite time, an idealization impossible under real life human conditions. The industrial society strives for growth, and increased transformation pace will accordingly imply both reduced efficiencies and larger generations of disorder (Høyer, 1979; Hall *et al.*, 1992).

The more controversial contribution in the Georgescu-Roegen theory is the claim that the second law fully applies to both matter and energy, and that it is completely valid for the earth-sun system understood as an isolated system (it is not a closed system as the exchange of sun energy is included, and it is not an open system as it does not include any matter exchange with outer space of any real interest). This is the basis for his *fourth law of thermodynamics*, stated as follows: 'A. Unavailable matter cannot be recycled; B. A closed system (i.e. a system that cannot exchange matter with the environment) cannot perform work indefinitely at a constant rate' (Georgescu-Roegen, 1981, p. 265). This proclaims for matter what the entropy law claims for energy. The claim is that in an isolated system, instead of tending towards heat death, when all energy is unavailable, it tends towards chaos, when all matter-energy is unavailable. Georgescu-Roegen's term 'fourth law' was not the best choice, as there is quite another fourth law of basic thermodynamics together with a third law. In line with critical realist ontology we get closer to reality by considering the laws of thermodynamics, also the Georgescu-Roegen fourth law, as *mechanisms*. They are definitely present in real life, but are moderated through their co-existence with other physical laws and phenomena, just as in the case of the law of gravitation which only manifests itself without moderations in idealized isolated systems. The seemingly counter-entropic process of photosynthesis and plant growth represents such moderation, still without an ensuing absolute violation of the laws of thermodynamics. In complex systems there are always differences between moderation and violation. Taking the photosynthesis moderation into full consideration would imply an adaptation of all human life both to the pace and volume of this process, an extensive retreatment from today's processes in industrial societies. This is what Georgescu-Roegen termed a *bioeconomy*. The Georgescu-Roegen critique of neoclassical economics is that it fundamentally violates all the laws of thermodynamics, not only his fourth law. Both economic growth as such and the claims of continuous recycling are *theorems of impossibility*. Even the *steady-state economy* launched by the American ecological economist Herman E. Daly (1972) is an impossibility. The only long-term possibility lies in human retreatment to bioeconomics, a *degrowth* (Georgescu-Roegen, 1981).

The issue of timescale is often raised in this discourse. If the whole issue is about a timescale of the sun ultimately burning down – about four billion years – so why bother. This is a misunderstanding of the complete message (Georgescu-Roegen, 1981, pp. 257–69). The Georgescu-Roegen critique is most fundamentally about the limited life of the industrial growth society – actually the Sigmund Kvaløy term. Within this global industrial growth society the impossibility may be a matter of a few centuries, even when there is extensive material recycling. But with a fundamental retreatment to the bioeconomy, based on the volume and pace of global photosynthesis, the human species can have a life of eternity. Recycling of matter may be a part of that life, but then only as local cycles not of molecules but of matter in useful shape, used ploughs, glass, and worn out metal tools.

Within neoclassical teaching literature it is common to present the economic process of the industrial growth society as a circular diagram, a pendulum movement between production and consumption within an isolated system. According to Georgescu-Roegen (1971) no due weight is given to the simple fact that there are inter-relations between the economic process as such and its material and physical environment, and that this encompasses human historical footprints. The same applies to Marxist economic theory, in which the system of reproduction is represented as an economic process completely circular and self-upholding; nature is considered as a free gift to humans, without any inherent value. The implications of the laws of thermodynamics are to the very opposite. Economical processes activate energy and matter resources. This activation implies the utilization of the nature stores and funds of low entropy, the high quality natural resources. What is generated is high entropy – disorder – in one form of another, at some place, and at some time. All real economic processes utilize the difference in entropy between the natural contributions of low entropy and the resulting generation of high entropy. Any economic process, as any physical process of transformation, is thus upholding itself through the tapping of low entropy. Through application of technological filters the tapping can be made more or less efficient, leading to larger or smaller resource use. But without the continuous input of low entropy from nature there are no future conditions for a continuous industrial production. This is not a circular process, neither is it a to and fro process. It is a one-way process as regards loss of low entropy (Høyer, 1995).

According to the Georgescu-Roegen theory any economic process in industrial societies, recycling or not, is a process of loss in a thermodynamic sense. Larger economic volumes, larger growth, and thus larger paces of the recycles one should try to establish, only increase the losses. He thus presents a harsh critique of value theory both in neoclassical and Marxist economy. Only 'things' with low entropy can have value. High entropy is valueless or has negative value. Low entropy represents the necessary condition for 'things' to get value. But they only get their economic value when the low entropic resources are transformed by a number of agents in the form of human labour and capital equipment. This transformation gives, according to Georgescu-Roegen, the sufficient condition. However, the Georgescu-Roegen emphasis is that the necessary condition for all economic value lies with the natural resources in themselves, and not in the human-made transformation of these resources. And that these conditions are continually worsened, for ever, the larger the volumes and growth of the economic process (Georgescu-Roegen, 1971; Høyer, 1995).

The Nordic ecophilosophers on growth: Arne Næss

In the book presenting the most comprehensive outline of Arne Næss' ecosophy in English, *Ecology, Community and Lifestyle* (1989), a general scepticism to growth in Gross National Product as a political goal is displayed. In chapter 1 of the book, the deep ecological movement and eight formulations of a 'deep ecology platform' are presented. Although economic growth is not mentioned explicitly in any of the eight points, the view that continual economic growth in industrial states is incompatible with development based on deep ecological principles is displayed in the explanatory remarks of point no. 6. This point reads (Næss, 1989, p. 29): 'Significant change of life conditions for the better requires change in policies. These affect basic economic, technological, and ideological structures.' While this formulation in itself may appear somewhat vague in its wording, the explanatory remarks following immediately after the eight points leaves little doubt (ibid., p. 31): 'Economic growth as conceived and implemented today by the industrial states is incompatible with points (1) to (5).'

Although Næss has internationally often been associated precisely with the formulations of the platform for the deep ecology movement, these formulations may not be the most relevant place to look for Næss' opinions on the growth issue. The platform formulations were co-authored with the American George Sessions and were meant as basic views held in common by supporters of the deep ecology movement. Different deep ecologists may have more elaborate and specific opinions on particular eco-political issues, and Næss himself is of course no exception. One of the chapters of *Ecology, Community and Lifestyle* is dedicated to a critique of the prevailing view on economic growth within the discipline of economics. Here, Næss criticizes the widespread tendency among politicians and the media to refer to the Gross National Product (GNP) as if it were an indicator of progress. Næss argues for ignoring GNP in the industrial countries, among other things because GNP growth supports irresponsible and unsolidaric resource consumption and global pollution; favours wants instead of needs; favours hard and distant technologies; discriminates against people working at home; and has no definite correlation with the level of employment (Næss, 1989, pp. 113–16). According to Næss:

> practically everything signifying progress towards sane, ecosophically justifiable life, may result in 'lamentable' decrease in GNP and will be noted as 'worrying' decline in economic growth. Ecology is suspect . . . Life quality itself is suspect. Every use of anti-depression pills is a plus in GNP.
>
> *(ibid., p. 113)*

On the other hand, Næss is reluctant to explicitly recommending zero-growth or degrowth. He writes that there is no simple answer to the question of whether GNP should, within an ecologically responsible policy, grow rapidly, show decreasing rate of growth but still grow, remain stationary, or decrease. Instead, he characterizes GNP as irrelevant (ibid., p. 114). Actually, he describes the notion of zero-growth as a pejorative label invented by growth proponents: 'It is the defenders of the actual growth increase who have postulated fictional zero-growth as comfortable but non-existing opponents.'

Although Næss clearly considers the policies pursued in industrial countries to promote GNP growth to be detrimental to deep ecology goals, he thus seems unwilling to rule out the possibility that GNP growth based on a different composition of entries in the GNP could be compatible with those goals. In another text (Næss, 1993), the eight points of the deep ecology platform are again presented (here referred to as 'basic principles'), but the comments on the various points have been to some extent revised. This also applies to the explanatory remarks on point no. 6, where, in addition to the sentence already quoted above, the following words have been inserted immediately after (ibid., p. 413): 'There is only a faint resemblance between ideal sustainable forms of economic growth and the present policies of industrial societies.' At least, this formulation introduces some ambiguity as to whether Næss could be considered to represent a degrowth perspective or simply a more radical and ecocentric position within ecological modernization. This interpretation of Næss as not being fundamentally against economic growth is also expressed in David Rothenberg's introductory chapter of *Ecology, Community and Lifestyle*, where he writes (Næss, 1989, p. 17): 'Perhaps economic growth for its own sake is rejected, but not growth or progress defined in a more ecological manner.'

While the desirability of a decrease in the human population is the explicit topic of one of the eight basic deep ecological principles, criticism against economic growth only appears, as mentioned above, in the explanatory remarks, and with a less clear statement in the 1993 article than in the 1989 book. Possibly, this might reflect a wish not to scare away potential supporters of the deep ecology movement. At several occasions in *Ecology, Community and Lifestyle*, Næss warns deep ecologists against being sectarian, and he emphasizes the need for environmentalists to establish acquaintance with 'green economic classics' and the relevance of having some supporters of the deep ecological movement who are able 'to point out what is the best solution when decisions are made in economic terms' (Næss, 1989, p. 106). In order for deep ecology supporters not to be branded as 'extreme' and excluded from such arenas, Næss may have found it opportune not to flag a total rejection of economic growth.

Be it for tactical reasons or as a reflection of his actual conviction, Næss' growth criticism as displayed in his main writings must be characterized as conditional rather than categorical, even when considering only growth in the industrial societies. The present conditions are, however, fundamentally different from the conditions under which economic growth in industrial countries could hypothetically be evaluated positively from a deep ecological standpoint. Anyway, Næss' arguments for the intrinsic value of non-human life, for humanity's moral responsibility towards non-human nature, and for quality of life as something fundamentally different from consumption, form important potential building blocks in a line of argument for degrowth, when combined with statements about the implausibility of achieving a sufficiently high degree of decoupling between growth and negative environmental impacts to offset the eco-destructive effects of economic growth.

In the English – or rather American – 1989 edition of *Ecology, Community and Lifestyle* there are several literature references to works by Georgescu-Roegen. These references were actually not at all present in the Norwegian original from 1974, when Georgescu-Roegen's works were much highlighted within the Norwegian ecophilosophical community. However, the basics of the Georgescu-Roegen growth critique is still not covered, or accepted. To apply the terms of Næss himself; his growth critique belongs to *shallow ecology* and not to his preferred deep ecology platform.

As mentioned above, Næss states (in one of the eight deep-ecological principle statements as well as elsewhere) that policy changes affecting basic economic, technological and ideological structures are necessary to obtain a development in accordance with deep ecology principles. In *Ecology, Community and Lifestyle*, societal conditions (economic, technological and institutional) as well as individual behaviour and attitudes (patterns of consumption, an instrumental view on nature) are pointed at as causes of environmental problems. He is, however, not very specific in explaining why these circumstances work the way they do. Apparently,

Næss sees a main cause of the environmental problems in the fact that most people have dualistic conception of human and non-human nature, and that most people do not to a sufficient degree question whether concrete decisions are in accordance with fundamental objectives and values. He is, however, not very specific in his critique of the societal conditions. His criticism against GNP growth as a societal goal is, for example, not followed by any in-depth analysis of how certain traits of the economic system create an urge for continual growth.

Instead, in the book's chapter 6 (on eco-politics), Næss states that it is essential for proponents of green politics to insist and show that they cannot be placed along a scale from red to blue. He also provides a diagram where the 'greenness' converges towards an axis midway between the blue and the red political pole. As examples of similarities between red and green, Næss mentions the stressing of social responsibility, very high priority of fighting undesirable ethical, social and cultural consequences of the unrestrained market economy, equality and opposition to hierarchical structures. As examples of similarities between green and blue, he mentions the stressing of the value of personal enterprise (overlapping the blue private enterprise) and very high priority of fighting bureaucracy (Næss, 1989, p. 13).

Næss does, however, not mention the growth imperative of the capitalist economic system. Although not discussing the matter explicitly, it may seem as though he assumes that if present big business is replaced by some sort of small-scale capitalism (cf. his valuing of personal enterprise), this will result in a steady-state economy. He does, however, not discuss what regulations would be necessary to avoid the sum of profit-making strategies of all the small-scale capitalists to result in aggregate-level growth, nor the regulations necessary to replace multinational corporations with small, local and environmentally friendly enterprises.

Sigmund Kvaløy on growth

Whereas Arne Næss, in spite of his strong scepticism to GNP as a measure of progress, did not rule out the possibility of a radically different, ideal form of sustainable economic growth, Sigmund Kvaløy's texts are more categorically rejecting economic growth in industrial countries. According to Kvaløy, 'Industrial Growth Society' is the antidote to his ideal of a 'life necessity society'. In his main work *Økofilosofisk Fragment*, Kvaløy (1974) states that the primary characteristic of the contemporary eco-crisis is a 'growth mechanics' that continuously 'simplifies' ecosystems, not only locally but at a global scale, as a sort of 'world-wide crippling growth' (ibid., p. 10). According to Kvaløy, humans cannot avoid to erode the diversity of life on earth ('the life community', as he terms it) to some extent. The point is, however, that the speed of this erosion now increases year by year. The accelerating growth in the erosion of the diversity of life is caused by exponential economic growth and can be observed in terms of oil extraction, hydro-electrical development, the degree of pollution, the speed of urbanization, etc. (ibid., p. 12). Kvaløy mentions that a similar growth also can be observed in the size of the global population. His emphasis is, however, clearly more on economic growth than on population growth. Here, an important difference can be seen, compared to Arne Næss' deep ecology platform, where population growth is mentioned explicitly and economic growth only in the comments on one of the platform formulations.

In this there are many connections to the theorizing of Georgescu-Roegen. And Kvaløy himself claims to be extensively inspired by Georgescu-Roegen; 'Georgescu-Roegen is one of the most crucial contributors to the development of the particular Norwegian ecophilosophy' (personal communication, April 2011). In his *Ecophilosophical Fragment* from 1974, Kvaløy thus emphasizes: 'The Georgescu-Roegen book *The Entropy Law and the Economic Process* is a book that is a rich treasure-chest for all those environmentalists who are frustrated from beating in vain on the fortress walls of the economists' establishment' (Kvaløy, 1974, p. 155).

According to Kvaløy, managing the development in a way preventing the eco-crisis from turning into eco-catastrophe might perhaps not be such a tremendously difficult task if growth took place only within one or a few areas. However, since growth occurs within a large number of fields simultaneously, it is extremely difficult to realize its consequences at a sufficiently early stage. Kvaløy gives a colourful illustration of a growth game with eight players located in different Nordic cities, each given the task of putting a number of game pieces (symbolizing factories, airports, bridges, harbors, oil drilling platforms, tankers, etc.) on a playing board designed as a map of the Nordic countries. The first five minutes of the game, each player has to locate a certain number of game pieces on the board, the next five minutes, the same amount plus a certain increase, and so on. As the game goes on, the speed of locating new pieces on the map increases to such an extent that the players have no longer time to communicate with each other, and soon the players have to crawl up on the map to find vacant spots to locate their expanding enterprises. The players do not become aware of each other until the end of the game, where they meet – completely exhausted – in the middle of the map. At that time, there are no longer any green or ocean-blue spots left on the map (Kvaløy, 1974, p. 14).

As illustrated in the game metaphor, Kvaløy sees 'industrial growth society' as destructive to nature as well as alienating to humans. It is a system in crisis, on its way toward eco-catastrophe if allowed to continue on its path. Kvaløy defines the eco-crisis as a 'simplification' (i.e. a reduction in the diversity) of the global ecosystem, caused by a requirement for ever increasing profit and an acceleration of the development within a very wide and continually increasing number of fields (ibid., p. 14). Notwithstanding Kvaløy's deep worry about the impacts on non-human nature, he considers that the most important aspect of the eco-crisis is the changes its driving forces impose on the human society. Here, too, a 'simplification' takes place at accelerating speed, making it more and more difficult for humans to become aware of what happens and change to a different way of life (ibid., p. 15).

Kvaløy explicitly dissociates himself from the view that increased economic growth is necessary in order to be able to afford environmental protection. He thus rejects a line of argument later to be known as the hypothesis of the 'Environmental Kuznets curve' (and also making up part of the argumentation of the World Commission on Environment and Development (1987) for 'reviving economic growth').

Whereas Arne Næss is silent about capitalism's growth imperative and tends to emphasize changes in individual lifestyles and local community initiatives (cf. also the title of his book *Ecology, Community and Lifestyle*), Kvaløy is explicit on the role of the overall economic-structural conditions in society. According to Kvaløy, 'there are certain indications that our modern industrial society would break down into chaos if it is not allowed to consume resources and change life-communities at accelerating speed. Even a *steady* growth cannot be tolerated without breakdown' (Kvaløy, 1974, p. 15, see also p. 90).

In a separate chapter, Kvaløy elaborates on the tendency of 'Industrial Growth Society' to require economic growth at accelerating speed. According to Kvaløy, a key driving force in this society is individual economic competition, partly between industrial enterprises, partially between the society as a state and other states in the world. But there are also, according to Kvaløy, two other key features of Industrial Growth Society pushing it along its growth trajectory: quantified planning (leading to 'erroneous signals' to the managers due to omission of crucial, but non-quantifiable conditions and impacts), and applied knowledge in technology development. Kvaløy considers that none of the three features can be removed without a fundamental and radical change of society as a whole (ibid., p. 91).

Kvaløy stresses the interwoven nature of the three features, which are claimed to have the same historical roots. He therefore explicitly distances himself from the opinion that only the first of the three (economic competition) should be removed while allowing the two latter (quantification and technological expansion) to remain (ibid., p. 91). In line with this, Kvaløy makes clear that his terms 'Industrial Growth Society' and 'Techno-capitalism' do not only refer to West Europe/USA/Japan, but also to the system in the Soviet Union, which is characterized as a sort of 'giant monopoly' based on similar individualistic material power struggles and engaging in the global competition in similar ways as the Western states (ibid., pp. 90 and 96).

In a more recent article, Kvaløy (1991) uses a more poetical language to describe the crisis-prone Industrial Growth Society and its driving forces:

> It is impossible to prevent loss of the water in a boiling pot by fixing the lid by means of bolts – not as long as the heat underneath is on . . . In particular, the situation turns exciting since all the [three or four main global] pressure-boilers are dependent on the same wood, oil, rain forests, etc. The excitement reaches its climax when it is discovered that the boilers are fed by robot systems unable to do anything but heating. For they have been constructed in accordance with a worldview saying that competition is the fundamental need. And that those who do not compete shall die.
>
> *(Kvaløy, 1991, p. 35)*

According to Kvaløy, the obvious alternative to Industrial Growth Society would be 'the balanced life-necessity society', i.e. a society based not on the production of industrial artifacts but on the production of food and other mere life necessities. In this society, industry would mainly serve the production of food, clothing and fuel. Such a basis for society would, among other things, entail a strong reduction of the global division of labour, accompanied by increased local self-sufficiency and a transition to solar energy as the prime energy source (ibid., p. 91). Clearly, what Kvaløy envisages here is not only a zero-growth society, but a society with clearly lower volume of economic activity than what characterized the industrial societies when the book was written. Given the fact that economic growth has proceeded during nearly four decades since then, it is clear that Kvaløy's position in relation to the contemporary growth debates is one of degrowth, as distinct from steady-state economy based on current, but non-growing volumes of production and consumption. In this he definitely demonstrates his close connection to the Georgescu-Roegen theories.

Torsten Hägerstrand on growth

In several articles, Torsten Hägerstrand maintains that there are limits to economic growth. In particular, this perspective is more and more evident in his latest books and articles. But his growth criticism, emphasizing environmental as well as social consequences of growth, is also expressed in some of his writings from the mid-1970s.

In a lecture held in 1984 and published seven year later, Hägerstrand (1991) calls attention to the limits to exponential growth in throughput of material substances in the economic system. According to Hägerstrand, the Earth is too confined to ever make it possible for humanity on all continents to live with the commodities and habits that Swedes have become accustomed to. Hägerstrand still believes that quantitative scarcity seriously curbing the growth in the industrial civilization may not appear until a couple of centuries from now.

However, social limits to growth will, according to Hägerstrand, appear much earlier. Referring to the economist Fred Hirsch, Hägerstrand holds that there will be more congestion, queues and dissatisfaction as more and more people aspire to reach the, after all, relatively few attractive positions and positional goods. Notably, the available time sets limits. It will be difficult for people to spend sufficient time on income earning while also having time to make use of the rising number of commodities and engage in other kinds of attractive consumption (ibid., p. 91). Hägerstrand points at an inherent contradiction between rising productivity in the production of commodities and the much lower potential for productivity gains within many times of services: it takes time to teach a child to read, and a symphony cannot be improved by speeding up the pace of drumming. The information technology holds a similar contradiction: mass distribution of information is easy, but each receiver can only absorb a limited part at a time (ibid., p. 92). And in parallel with growth in the amount of material devices and commodities, information, and scientific and technical knowledge, a growing number of rules and regulations are established as fences around our lives (ibid., p. 189).

According to Hägerstrand, some of the industrial countries have reached a 'saturation level' where the material needs of most people have been met (or even overprovided). On the other hand, social, emotional and aesthetical needs have been neglected. New commodities contribute to a lesser and lesser degree to meet real needs (ibid., p. 92). Here, Hägerstrand's argumentation resonates with the considerations of scholars like Easterlin (1973) and Jackson (2009) who hold that further economic growth contributes little or not at all to human health, life satisfaction and well-being once a threshold level of affluence has been reached.

According to Hägerstrand, the 'invasion of technology' and the increased specialization of labour have caused a number of side-effects which have, in their turn, given rise to new technical and maybe also bureaucratic measures to deal with these negative consequences. These side-effects can be seen as a sort of 'friction'. In the high-specialized society, the number of frictions has reached a level where, according to Hägerstrand (ibid., p. 190), 'it is, as it were, like some abstract container has been filled to the brim with components and component systems that are each, seen in isolation, completely rational, but which together are now in all kinds of friction with each other'. The progress in specialization and division of labour has disregarded the fact that some entities have a limited capacity for being divided and manipulated, writes Hägerstrand. 'On the one hand we have the indivisible individual who will not or cannot subdivide herself within any kind of artificial and disjointed pattern of roles and tasks. On the other hand we have another whole, namely the biophysical population/environmental system, which in a similar way cannot sustain any unlimited amount of technical and organizational stress' (ibid., p. 190).

Hägerstrand is in most of his writings not very specific about the deeper social causes of the 'technological invasion' and the environmentally and socially disruptive growth in production and consumption. Most often, his formulations are of a more subtle nature, hinting at the driving forces rather than mentioning them explicitly. In an article written in 1989, he points at 'the tyranny of small decisions' and 'the inheritance from Decartes' as important contributors to the global ecological crisis. The term 'small decisions' clearly hints at decentralized market-based decisions absent from coordination by any comprehensive regulatory body. Hägerstrand is particularly worried about the aggregate consequences of the many fragmented micro-scale decisions determining technological development. Using the proliferation of automobiles as an example, Hägerstrand underlines the multi-faceted range of environmental impacts, of which the pollution created from its exhausts is but one category. As he sees it, one of the main problems of the decentralized and uncoordinated system of decision-making is that the negative environmental impacts of local actions are not visible to the local actor: 'there are no natural global mechanisms punishing the local actor if he is driving into a blind alley' (ibid., p. 193).

The liberal market, absent from strong regulations to protect global and local environmental qualities, is thus clearly a target of Hägerstrand's criticism. However, distinct from, e.g., Sigmund Kvaløy, Hägerstrand does not mention the growth imperative of the economic system per se. Instead, he seems to find the main driving force in technological development based on a Cartesian reductionist worldview. Descartes' ontology and methodological rules have led to an accelerated fragmentation of scientific knowledge, where there is a glaring disproportion between highly effective micro-knowledge and only vague or non-existing ideas at a macro level (ibid., p. 194). Quoting the ecologist P. Odum (1989), Hägerstrand points to the fact that the open systems in the real world cannot be enclosed in glass tubes or laboratory walls. Indirectly, Hägerstrand here identifies lack of critical realist ontological understanding (i.e. of the world as complex, open-systemic, stratified and multi-causal) as an important driving force leading to eco-destructive technological development and failure to realize the limits of economic growth.

Georg Henrik von Wright on growth

While Georg Henrik von Wright is perhaps most known internationally for his writings on analytic philosophy and philosophical logic, he has since the late 1980s also written a number of critical texts about the 'techno system' and myths of progress based on technological development and bureaucratic-paternalistic kinds of social engineering. Especially in one of his essays, 'A Provocative Pessimism', published as postscript of his 1993 Swedish-language book *The Myth of Progress* (and reprinted as Chapter 8 in this book), von Wright displays a strong criticism of the pursuit of technological development mainly as an instrument to promote economic growth (von Wright, 1993). Partly, this criticism is based on a disbelief in the possibility of predicting and controlling the side-effects of new technologies, also expressed in his book (Swedish-language) *Science and Reason* (1986). Here, von Wright draws on developments within 'chaos theory' and research on ecosystems as well as theoretical physics to reject the possibility of predicting in a deterministic way the impacts of technological intrusions into natural systems. He criticizes the blind pursuit of instrumental rationality in technological development where the rationality of the purposes that the technology aims to serve is seldom discussed and questioned. This lack of value-rational discussion contributes to depict technology-aided economic growth as an unquestioned manifestation of progress. Von Wright's arguments here show clear resemblance with those of Arne Næss (1989).

However, he also develops a much more fundamental and pessimistic standpoint, very much in line with the Zapffe philosophy of the tragic human species. In the concluding chapter of his book on *Science and Reason* (1986, p. 151) he emphasizes:

> A perspective which I don't consider unrealistic is that humans develop towards extinction as a zoological species. This is a perspective which many a time through history has stirred human societies. But I believe that it has a more pressing background now than ever before. For my own sake I can't find it very upsetting. Surely, one time or another the human species will

> no longer exist, whether it happens after a few hundred thousand years or after a couple of hundreds of years, this is of minor interest in relation to the ultimate cosmic perspective. When we consider all the other natural species the human race has led into extinction, such a nemesis of nature can seem quite fair.
>
> *(Our translation)*

According to von Wright, economic growth has become a main motivation for the expansion of the techno-system. As a result, scientific progress and accumulation of knowledge tend to be identified with economic growth per se. Von Wright considers this identification as a sort of reification and quantification of progress, stripped of any value content (von Wright, 1993, p. 42).

Von Wright admits that the increased material affluence brought about through industrial and technological development has in many cases entailed real progress in terms of improved conditions for human life. However, this does not, he writes, mean that growth still represents progress when the level of affluence has reached a certain level, or when its repercussions on the environment or the social order are taken into consideration (von Wright, 1993, p. 40).

Although von Wright does not elaborate on human's ethical obligations to nature, he mentions gene technology, animals rights, and the possible intrinsic value attached to species of plants and animals and landscape types as important ethical issues. Von Wright's arguments about environmental limits to growth are, however, mainly based on anthropocentric premises. Von Wright argues (1993, p. 49) for an expansion of solidarity from encompassing those similar to ourselves to a global responsibility also including future generations of humans. He does, however, in this context not mention non-human nature.

Von Wright anyhow finds solid anthropocentric arguments for his growth scepticism. He writes that he does not know of any convincing or even serious argument for the possibility of continual economic growth (ibid., p. 138). He criticizes the concept of sustainable development promoted by the World Commission on Environment and Development (1987) for containing a fundamental contradiction between, on the one hand, a realization of the alarming degradation of natural systems and an aim of reducing this degradation through energy saving, reduction of unnecessary consumption, etc., and on the other hand a commitment to continual economic growth not only in poor countries, but also in industrial countries. Von Wright specifically rejects the notion that growth in the rich parts of the world is necessary in order to enable us to help the poor countries to improve their conditions. Instead, von Wright concludes that the contradiction inherent in the concept of sustainable development 'can only be resolved through a movement in the opposite direction of the present, expansive, growth-promoting tendency. It is not only a matter of stopping the explosive population growth. There is also a call for what could be called a downscaling or cooling-down of the global economy' (von Wright, 1993, p. 139).

Von Wright considers the above-mentioned standpoint as rather trivial. However, in spite of the intuitive reasonableness of the need to put an end to economic growth, politicians as well as populations seem to be paralyzed, resorting to a false hope for salvation through technical fixes. In some arenas, resolutions on the need to save the environment are formulated, but in other, more powerful arenas, policies giving priority to 'stimulate the economy', reduced taxes, higher wages, and improvement of companies' international competitiveness are adopted (ibid., p. 143). According to von Wright, putting an end to economic growth would require strong regulation of market mechanisms, presumably by the state. This would presuppose that the state possessed great political authority, compared to the economic system. However, the current political-economic development proceeds in the opposite direction, with the European Union as a forerunner for deregulation and the subordination of the political system under the economic system. Von Wright is therefore pessimistic about the possibility of avoiding an accelerating deterioration of the environment and human suffering before the economic growth finally ends because it can no longer be sustained by its natural resource base. His pessimism is, however, not one of paralysis. Instead, he wants to be a 'provocative pessimist' who arouses people to reflect on and act to overcome the conditions now preventing the realization of a truly sustainable development.

Arne Johan Vetlesen on growth

During recent years, Arne Johan Vetlesen has been one of the most profiled critics of economic growth and its negative environmental impacts in the Norwegian intellectual debates. In a number of newspaper feature articles, he has identified economic growth, consumerism and capitalism as key causes of ecological unsustainability in general and the climate crisis in particular. However, in his scholarly writings as a philosopher, where he has been active for three decades, economic growth has not been addressed until recently. However, in a Norwegian-language anthology published in 2008 entitled *Nytt klima* (New Climate) he presents a number of challenges to environmental ethics entailing an sharp criticism of economic growth as a societal objective, based on considerations about the value we attach to nature, a critical view on technology development, and the rapid worldwide destruction of non-human nature. Vetlesen's growth criticism is in particular displayed in the final part of his extensive main chapter of the book, but also in a short introductory chapter.

Similar to Arne Næss, Vetlesen (2008a, p. 14, 2008b, pp. 114–15) calls attention to the many points of criticism that have been raised by critics within as well as outside the discipline of economics against the use of GNP as an indicator of human progress. Vetlesen points to the shallowness of the values underlying the conflation of GNP growth with increasing prosperity: happiness is glued to having, not being, and to spend, not to care about or protect. According to Vetlesen, this set of values aspires to domination in nearly every aspect of life. There is no standard for limits, and no conception of sufficiency, or that a goal has been reached in the sense that enough is enough. Expansion has become a goal in itself, as has an increased assortment of commodities. The wheels must be kept rolling, at any cost.

And this is the crucial issue, according to Vetlesen: the growth has its huge costs. The expansion now threatens the viability of the planet. What is called for today is therefore moderation, not for fun, but to maintain an ecologically intact planet (Vetlesen, 2008a, p. 14).

According to Vetlesen, there is 'a fundamental incompatibility between [on the one hand] an economy explicitly and essentially geared toward continual and non-terminable growth, and [on the other hand] a planet containing only a limited amount of vital natural resources (sources of food production and energy), where this very planet is identical to the single planet on which the formerly mentioned economic activities are bound to take place' (Vetlesen, 2008b, p. 115). Whereas the notion of 'natural resources' in this quote could possibly be interpreted as a concern for nature only for its instrumental value for human purposes, Vetlesen makes it very clear in the remaining text of his main chapter of the book that he considers humans to have moral obligations to the rest of nature. The threats in the form of reduced biodiversity and extinction of species are thus, in Vetlesen's view, as important impacts of climate change and the ecological crisis as are their impacts on the availability of resources for human purposes. According to Vetlesen, an economic system oriented towards growth and increased affluence is a hyper-efficient killing machine in terms of extinction of natural species (Vetlesen, 2008a, p. 12). Vetlesen points to the fact that it is not possible in the long run to pursue the realization of limitless and infinite growth on a planet which, from nature, is and continues to be a limited, finite universe (Vetlesen, 2008b, p. 115). Here, the resemblance between Vetlesen's arguments and the ecological economist Herman Daly's (1993) 'impossibility theorem' is obvious.

However, distinct from Daly, Vetlesen also claims a second impossibility: that of reconciling the capitalist system with a non-growing, environmentally sustainable economy. According to Vetlesen, 'the key characteristics of a capitalist economy, (i.e. its system-immanent twin objectives of growth and profit) prove to be exactly the traits resulting in its actual incompatibility with the planetary key characteristics of limits and finiteness' (Vetlesen, 2008b, p. 115). In the era of globalized capitalism, all human needs are directed towards the market, and all the cultures of the world are forced to strive for the very instrument – the market – which continuously robs humans of their ability to survive by their own efforts and skills and to live lives that do not in total lead to an overload of the global carrying capacity – the limited which is threatened by the limitlessness of the market and the insatiableness of the commercialized needs (ibid., p. 102). These dynamics lead Vetlesen to the conclusion that 'those who talk about the climate crisis cannot be silent about capitalism' (Vetlesen, 2008a, p. 11): the logic of the capitalist world economy is the logic of limitlessness, whereas the logic of the planet is one of limits and dependency (ibid., p. 14). In a situation where most of the Norwegian environmental NGOs (including the previously rather activist youth organization of Friends of the Earth Norway) are reluctant to use the word capitalism, Vetlesen encourages debaters to call a spade a spade and give some of the basic and structural causes of the environmental crisis their right names. He also warns against underestimating how strong and powerful are the interests champions of sustainability are up against: 'Strong economic and political interests are firmly determined to retain their power to navigate the planet toward the abyss, and humanity toward a warned collective breakdown' (ibid., p. 11).

Technologies enabling humans to make ever-increasing encroachments on nature and requiring more and more energy (although sometimes marketed as environmentally friendly and energy-efficient) are of course an important part of the picture. But according to Vetlesen, the development and environmental impacts of technologies cannot be regarded in isolation from their societal context. What we are facing is interplay between technologies and commercial interests. The technologies that are actually given priority are those serving to fulfil the goal of maximal profit within a limited time horizon (contemporary owners and consumers). One exploits what one is able to exploit (ibid., p. 101).

Our own work on growth and degrowth

The authors of the present chapter have for a long time been sympathizing with the Nordic ecophilosophical movement[1] and have also at some occasions engaged in academic writing on ecophilosophical issues (Høyer *et al.*, 1977; Høyer, 1979, 1995, 1999; Næss, P., 1992, 2006; Høyer and Næss, 2008; Næss and Høyer, 2009a, 2009b). An ecophilosophically informed normative standpoint has to a high extent influenced our choices of research topics within our academic fields of research (urban studies, energy, mobility, planning, and impact assessment). During the latest decade we have engaged in ontological and epistemological debates on the fundamental assumptions underlying research within these fields, informed by a critical realist perspective. A main issue in this endeavour has been to defend the possibilities of planning and critique of likely impact of urban and infrastructure development against the dominating scepticism and relativism of the postmodern era. Increasingly, we have become aware of the great similarities between Nordic ecophilosophical thought and key critical realist understandings (see Høyer and Næss, 2008). Partly, we have worked empirically exploring the possibilities and limits of ecological modernization, especially within the domains of urban development and mobility. However, building on Benton, Dunlap and Catton and others in formulating a critical realism-based ontology for cause-oriented studies of environmental problems, and a growth criticism rooted in Nordic deep ecology as well as eco-Marxist analyses (Kovel, O'Connor, Harvey, etc.), we have also engaged in theoretical and normative writing on the limits to growth and the social roots of the contemporary growth trajectory.

In line with Sigmund Kvaløy, Høyer has through more than thirty years contributed with ecophilosophical elaborations of the entropy theories of Georgescu-Roegen. One of his major contributions was a book published in 1979, *Oil, Energy and Planning* (Høyer, 1979) where he outlined the implications of thermodynamics and the wider Georgescu-Roegen theories for natural resources theory and husbandry. In a 1995 report for the Research Council of Norway this was developed into a more coherent theory of natural resources and economy (Høyer and Groven, 1995).

Distinct from Sigmund Kvaløy (1991), who tends to reject the notion of sustainable development as understood by the United Nation's World Commission on Environment and Development (1987), we have since the publication of the Commission's report *Our Common Future* generally accepted this concept and used it a basis for criticism of environmentally harmful and socially unjust kinds of development. In particular, we have supported the definition given in the beginning of Chapter 2 in *Our Common Future*, where it is stated that sustainability is about meeting the needs of present and future generations of humans, in particular the needs of the world's poor, to which overriding priority should be given (ibid., p. 43). There are also some passages in the report admitting that humans have moral responsibilities toward non-human nature, although the report also has several recommendations for policies that in their effect neglect these responsibilities as well as those towards future generations of humans. In our writings, we have criticized these recommendations for violating the values inherent in the concept of sustainable development as used by the World Commission itself, but also the widespread tendency of watering down the concept in much of the subsequent academic and political literature. In particular, the so-called economic dimension of sustainability has in the contemporary dominating discourse largely been re-interpreted as a goal of promoting traditional economic growth, whereas in *Our Common Future* economic growth was considered as a measure and not a goal in itself, and it was also stated that its content need to be changed.

The World Commission's conception of economic growth in a sustainability context was thus in line with the idea of ecological modernization, according to which economic growth must and can be 'decoupled' from negative environmental consequences, thus allowing growth to continue unlimited without jeopardizing environmental sustainability. In several publications, we have criticized the ontological, empirical and ethical assumptions underlying this belief, which we claim to be based on an 'autistic' neoclassical economic paradigm virtually immune to insights from other disciplines.

Real interdisciplinarity would instead most likely lead to the conclusion that the economic dimension of sustainability should – at least in rich countries – imply zero or negative growth, instead of the tacit subscription to the growth imperative characterizing most interpretations (Høyer and Næss, 2008). The 'decoupling' of growth from environmental impacts that would be required in order to combine long-term growth with environmental sustainability would presuppose an unrealistically high degree of eco-technological efficiency increase (Høyer and Næss, 2001; Næss, P., 2006; Næss and Høyer, 2009a) as well as an unrealistically high degree of control of the 'machines and their products' (Kvaløy, 1985) and their impacts on the surrounding ecology. Moreover, there is considerable evidence indicating that people will not become happier from consuming ever more commodities and services, once a threshold value has been reached (Easterlin, 1973; Jackson, 2009). Economic growth in rich countries is therefore hardly conducive to the quality of life of its own inhabitants, and it is clearly incompatible with meeting the needs of the world's poor and future generations of humans. Not the least is it sharply at odds with 'the flourishing of non-human life' (Næss, A., 1989).

Instead of relying on economic growth as a mean to meet the needs of the world's poor through the 'trickle-down effect', we argue for radical redistribution: redistribution of wealth from rich to poor population groups within each country, from rich to poor countries at a global scale, and redistribution of resource consumption from present to future generations. The radical redistribution should also include a redistribution of the hours of the day towards a lower number of hours spent on wage-labour and a higher number of hours spent on family and friends, culture, experiencing nature, and participation in social life (Næss and Høyer, 2009b).

In line with critical realist ontology as well as the type of interdisciplinary thinking that has been advocated for several decades by ecologists and ecophilosophers, we warn against 'disciplinary tunnel vision' (Næss, P., 2010) as a main obstacle to an environmentally sustainable and socially just development. Such interdisciplinarity is of course necessary in order to avoid the possibility that measures aimed at solving one environmental problem do not cause new and often equally serious ecological and/or social problems, as is often the case with 'technical fixes'. But the scope must reach beyond a focus on relationships within non-human nature and the impacts of human technologies and activities on natural ecosystems. There is an urgent need to address the social conditions contributing to environmental degradation. A profound change in humans' view of nature away from the presently dominating strong anthropocentrism toward a more ecocentric view is part of such a transition, as stressed by Arne Næss (1989, 1993). However, in line with Sigmund Kvaløy (1973, 1991) and Arne Johan Vetlesen (2008a), we consider individual lifestyle change within the frames of a capitalist, competitive economic system to be insufficient. An analysis of the so-called institutional dimension of sustainability (Spangenberg, 2002), i.e. the frameworks for implementing policies necessary to achieve sustainable development, would lead, if based on a truly interdisciplinary analysis of the environmental, social and economic dimensions of sustainability, to the conclusion that the growth-oriented, market liberal economic system would need to be transcended. Several inherent causal mechanisms of the capitalist economic system together make economic growth a necessity for this system to maintain its stability (Marx, 1858/1973; Kovel, 2002). The 'boilers' are indeed fed by robot systems unable to do anything but heating (Kvaløy, 1991). If continual economic growth is ecologically unsustainable, then so is therefore continual capitalism too (Næss, P., 2006; Næss and Høyer, 2009a).

The belief in the possibility of obtaining a sufficiently high degree of 'decoupling' and 'eco-efficiency increase' to allow economic growth to continue over the next century or longer without causing global ecocatastrophe should appropriately be characterized as 'ideological', since the set of ideas is false and at the same time more or less necessary for those whose interests would have been threatened if the practices among the population had not been strongly influenced by this belief (Bhaskar, 1998). Our arguments against the possibility of an environmentally sustainable, continual economic growth point, therefore, at the nature-destroying character of capitalism itself, and call for a profound change not only in individual values and lifestyles, but also in basic socio-economic structures. Of course, there are strong mutual enforcing mechanisms between cultural values and social structures, so a change towards an eco-friendly and socially just society would have to involve a transformation of both.

While arguing for the necessity of replacing capitalism, in particular in its neoliberal, globalized version, with some sort of 'ecosocialism', we also recognize the long-standing neglect of ecological limits in most of the Marxist tradition (especially in the Marxist-Leninist variant). In line with Kvaløy, we consider ecological and social sustainability to require not only a removal of the profit-oriented and competition-based economic system, but also of technological expansionism and number-fetishist planning (in particular evident in Cost-Benefit Analysis) leading to erroneous signals to decision-makers due to omission of crucial, non-quantifiable conditions and impacts. Technocratic conceptions of 'optimal' nature management in a solar-based growth economy are still influential among left-wing environmentalists, and we have argued strongly against these ideas (Næss and Høyer, 2009a, 2009b).

In the rich corners of the world, ecological realities call for a replacement of economic growth with *degrowth* (and not only with a steady-state economy based on current consumption levels). Such a contraction of the economy must, however, be accompanied with redistribution of wealth if it is to be socially sustainable (Martinez-Alier, 2009). In a non-growing economy, any increase in the wealth of the richest implies a reduced affluence level among the rest of the population. In a degrowth economy, the potential for conflict over the distribution of resources will be even harder. Obviously, strong social institutions must be in place in order to manage the double challenge of reversing the growth trajectory of the economy while securing the quality of life for all members of society.

Within the field of transport and mobility, a degrowth society would probably mean that not only the use of individual motor vehicles would have to be reduced, but also the total amount of transport (Høyer, 1999). The 'space-time compression' (Harvey, 1990) would have to be reversed, at least as regards physical mobility, and the friction of distance 'reinvented' and increased. Of course, technologies for more environmentally friendly vehicles should at the same time be welcomed. The desirability of more environmentally favourable solutions also applies to the field of urban development. However, in a degrowth economy, the question of reducing the volume of the building stock and the size of the urbanized land will probably enter the agenda. The construction of new and more environmentally friendly buildings through densification must then be combined with the abolition of some of the environmentally least favourable built environments, such as the most car-dependent single-family home areas and office parks. Such dismantling of the most resource-consuming and unfavorably located built structures would at the same time open possibilities for nature regeneration and a larger coherence of the natural areas and landscapes surrounding the city (Næss, P., 2011).

Note

1 In the 1970s, especially Høyer but to some extent also P. Næss participated in the so-called Samarbeidsgruppene for natur- og miljøvern (snm), an ecophilosophically inspired grassroot movement initiated by Sigmund Kvaløy, where also Arne Næss was active. Høyer was an eager member in (snm)'s ecophilosophy group and also was co-author with Kvaløy and other (snm) activists on a book criticizing the fundamental characteristics of the European Union (or to be more correct: its predecessor EEC) from an eco-political standpoint. Høyer published ecophilosophically informed books on energy policy and was active in the campaign against nuclear power in Norway. P. Næss participated in the (snm) group on urban environmental issues but was generally less active than Høyer in the movement.

References

Bhaskar, R. (1998): *The Possibility of Naturalism: A Philosophical Critique of the Contemporary Human Sciences*. Third Edition. London and New York: Routledge.

Daly, H.E. (ed. 1972): *Toward a Steady-State Economy*. San Francisco: W.H.Freeman & Co.

Daly, H.E. (1993): 'Sustainable Growth: An Impossibility Theorem'. In Daly, H.E. and Townsend, K.N. (eds): *Valuing the Earth: Economics, Ecology, Ethics*, pp. 267–273. Cambridge, MA: Massachusetts Institute of Technology.

Easterlin, R.A. (1973): 'Does Money Buy Happiness?' *The Public Interest*, Vol. 30, pp. 3–10.

Easterlin, R.A. (1995): 'Will Raising the Incomes of all Increase the Happiness of all?'. *Journal of Economic Behavior and Organization*, Vol. 27, pp. 35–47.

Georgescu-Roegen, N. (1971): *The Entropy Law and the Economic Process*. Cambridge, MA: Harvard University Press.

Georgescu-Roegen, N. (1976): *Energy and Economic Myths*. New York: Pergamon Press, Inc.

Georgescu-Roegen, N. (1979): *Demain la décroissance*, ed. by J. Grinevald and I. Rens. Lausanne, France: Ed. Pierre-Marcel Favre.

Georgescu-Roegen, N. (1981): Afterword, in: Rifkin, J. *Entropy. A New World View*. New York: Bantam Books.

Gorz, A., (1975): *Ecologie et Politique*. Paris: Le Seuil.

Hägerstrand, T. (1991): *Om tidens vidd och tingens ordning – Texter av Torsten Hägerstrand*. (On the Width of Time and the Order of Things – Texts by Torsten Hägerstrand.) Edited by G. Carlestam and B. Sollbe. Stockholm: Statens Råd för Byggnadsforskning.

Hall, C.A.S., Cleveland, C.J. and Kaufmann, R. (1992): *Energy and Resource Quality. The Ecology of the Economic Process*. Colorado: University Press of Colorado

Harvey, D. (1990): *The Condition of Postmodernity: An Enquiry Into the Origins of Cultural Change*. Cambridge, MA: Blackwell.

Høyer, K.G. (1979): *Olje, Energi og Planlegging* (Oil, Energy and Planning). Oslo: University Publishers.

Høyer, K.G. (1995): *By og land – hand i ratt. Mobilitet, miljø og bygd*. (Town and Country – Hand on the Steering Wheel. Mobility, Environment and Rural Communities). Report 4/95. Sogndal: Western Norway Research Institute.

Høyer, K.G. (1999): *Sustainable Mobility – the Concept and its Implications*. PhD thesis. Roskilde/Sogndal: Roskilde University and Western Norway Research Institute.

Høyer, K.G. and Groven, K. (1995): *Fisk og Miljø. Hvordan står det til med bærekraften?* (Fish and the Environment. How is it with Sustainability?). Report 5/95. Sogndal, Norway: Western Norway Research Institute.

Høyer, K.G. and Næss, P. (2001): 'The Ecological Traces of Growth: Economic Growth,

Liberalization, Increased Consumption – and Sustainable Urban Development?'. *Journal of Environmental Policy and Planning*, Vol. 3, pp. 177–92.

Høyer, K.G. and Næss, P. (2008): 'Interdisciplinarity, Ecology and Scientific Theory – The Case of Sustainable Urban Development'. *Journal of Critical Realism*, Vol. 7, pp. 5–33.

Høyer, K.G., Gundersen, H., Poleszynski, D. and Reinton, P.O. (1977): *Spillet om atom-kraften*. (The Play of Nuclear Power.) Oslo: Pax Publ.

Jackson, T. (2009): *Prosperity without Growth: Economics for a Finite Planet*. London: Earthscan.

Kovel, J. (2002): *The Enemy of Nature: The End of Capitalism or the End of the World?* Zed Books/Fernwood Publishing.

Kvaløy, S. (1973). *Økofilosofisk Fragment IV*. (Ecophilosophical Fragment IV). Trondheim: Tapir Publishers.

Kvaløy, S. (1985): 'Mangfold og tid. Hvordan en bryter pyramidens herredømme'. In Skønberg, S.E. (ed.): *Grønn pepper i turbinene – til ære for Sigmund Kvaløy Setereng. 16 bidrag om miljøkrisen og veiene ut av den*. Oslo: Universitetsforlaget.

Kvaløy Setereng, S. (1991): 'Systemkrise og krisesystem' (Systems Crisis and Crisis Systems). In Album, G., Alstadheim, K.B. and Gule, L. (eds): *Supermarked eller felles framtid. – EF, økokrisen og Norges valg*. Oslo: Cappelen.

Martinez-Alier, J. (2009): 'Socially Sustainable Economic De-growth'. *Development and Change*, Vol. 40, pp. 1099–119.

Martínez-Alier, J., Pascual, U., Vivien, F.-D. and Zaccai, E. (2010): 'Sustainable De-growth: Mapping the Context, Criticisms and Future Prospects of an Emergent Paradigm'. In: *Ecological Economics*, Vol. 69, pp. 1741–7.

Marx K. (1858/1973): *Foundations of the Critique of Political Economy*. Washington: Vintage Books, 1973 (originally published 1858 in German).

Meadows, D.H., Meadow, D.L., Randers, J. and Behrens, W.W. (1972): *The Limits to Growth* (Report to the Club of Rome). New York: Universe Books.

Næss, A. (1989): *Ecology, Community and Lifestyle*. Cambridge: Cambridge University Press.

Næss, A. (1993): 'The Deep Ecological Movement: Some Philopsophical Aspects'. In Armstrong, S.J. and Botzler, R.G. (eds): *Environmental Ethics: Divergence and Convergence*, pp. 411–21. New York: McGraw-Hill. (Reprinted as Chapter 6 in this book.)

Næss, P. (1992): *Miljøfilosofiske essays*. (Essays on Environmental Philosophy.) Working Paper 1992/115. Oslo: Norwegian Institute for Urban and Regional Research.

Næss, P. (2006): 'Unsustainable Growth, Unsustainable Capitalism'. *Journal of Critical Realism*, Vol. 5, pp. 197–227.

Næss, P. (2010): 'The Dangerous Climate of Disciplinary Tunnel Vision'. In Bhaskar, R., Frank, C., Høyer, K.G., Næss, P. and Parker, J. (eds): *Interdisciplinarity and Climate Change*, pp. 54–84. London: Routledge.

Næss, P. (2011): 'Economic Growth, Urban Development and Environ-mental Sustainability'. Forthcoming in Berutti, G. (ed.): *Inhabiting the Future*. Naples: University of Frederico II.

Næss, P. and Høyer, K.G. (2009a): 'The Emperor's Green Clothes: Growth, Decoupling and Capitalism'. *Capitalism, Nature, Socialism*, Vol. 20, pp. 74–95.

Næss, P. and Høyer, K.G. (2009b): 'Shadows in Schwartzmann's Sunny Society'. *Capitalism, Nature, Socialism*, Vol. 20, pp. 98–102.

Odum, E.P. (1989): 'Input Management of Production Systems'. *Science*, Vol. 243, pp. 171–82.

Spangenberg, J.H. (2002): 'Institutional Sustainability Indicators: An Analysis of the

Institutions in Agenda 21 and a Draft Set of Indicators for Monitoring their Effectivity'. *Sustainable Development*, Vol. 10, pp. 103–15.

Vetlesen, A.J. (2008a): 'Innledning' (Introduction.) In Vetlesen, A.J. (ed.): *Nytt klima. Miljøkrisen i samfunnskritisk lys*, pp. 9–20. (New Climate. The Environmental Crisis in a Perspective of Social Critique). Oslo: Gyldendal.

Vetlesen, A.J. (2008b): 'Natursyn, teknologi og miljøkrise – utfordringer for en miljøetikk' (View on Nature, Technology and Environmental Crisis – Challenges for an Environmental Ethics.) In Vetlesen, A.J. (ed.): *Nytt klima. Miljøkrisen i samfunnskritisk lys*, pp. 21–119. (New Climate. The Environmental Crisis in a Perspective of Social Critique). Oslo: Gyldendal.

Von Wright, G.H. (1986): *Vetenskapen och förnuftet.* (Science and Reason). Stockholm: MånPocket.

Von Wright, G.H. (1993): *Myten om fremskrittet. Tanker 1987–1992 med en intellektuell sjäselvbiografi.* (The Myth of Progress. Thoughts 1987–1992 with an Intellectual Self-Biography). Norwegian-language edition. Oslo: Cappelen.

World Commission on Environment and Development (1987): *Our Common Future.* Oxford and New York: Oxford University Press.

SELECTED TITLES ON NORDIC ECOPHILOSOPHY

Recommended further reading in English:

Former collections of articles

Henderson, B. and Vikander, N. (eds, 2007): *Nature First. Outdoor Life The Friluftsliv Way*. Toronto, Canada: Natural Heritage Books.

Reed, P. and Rothenburg, D. (eds, 1993): *Wisdom in the Open Air. The Norwegian Roots of Deep Ecology*. Minneapolis and London: University of Minnesota Press.

Sigmund Kvaløy Setereng

Kvaløy, S. (1974): 'Ecophilosophy and Ecopolitics: Thinking and Acting in Response to the Threats of Ecocatastrophe'. *North American Review* 259, no. 2, pp. 16–28.

Kvaløy, S. (1976): 'Survival Strength versus Life Strength – A Humano-Ecological System Perspective'. *Colloquium Internationale*, The International Organization for Human Ecology, Vol. 1, no. 6, Vienna.

Kvaløy, S. (1979): 'Man, Nature and Mechanistic Systems'. *North American Review* 264, no. 3, pp. 35–7.

Kvaløy, S. (1983): 'Let the Rivers Flow'. *Resurgence*, no. 96, pp. 23–6.

Kvaløy, S. (1988): 'Complexity and Time'. In: LaChapelle, D. (ed.) *Sacred Land*. Silverton, Colorado, US: Finn Hill Arts.

Torsten Hägerstrand

Hägerstrand, T. *et al.* (1974): *To Choose a Future*. (Translation from: SOU1972:59). Stockholm: Svenska Institutet.

Hägerstrand, T. (1977): *Culture and Ecology – Four Time-Geographic Essays*. Rapporter och Notiser 1977: 39. Lund, Sweden: Department of Social and Economic Geography, Lund University.

Hägerstrand, T. (1984): 'Presence and Absence. A Look at Conceptual Choices and Bodily Necessities'. *Regional Studies* 18, no. 5, pp. 373–9.
Hägerstrand, T. (1985): 'Time-Geography. Man, Society and Environment'. In: *The Science and Praxis of Complexity*, pp. 193–216. Tokyo: The United Nations University.
Hägerstrand, T. (1991): In: Carlestam, G. and Barbro Sollbe (eds) *Om Tidens Vidd och Tingens Ordning* (On the Space of Time and the Order of Things). Report T21:1991. Stockholm: Statens Råd för Byggnadsforskning. Several articles in English.

Arne Næss

Næss, A. (1973): 'The Shallow and the Deep, Long-Range Ecology Movement. A Summary'. *Inquiry* 16, pp. 95–100.
Næss, A. (1979): 'Self-Realization in Mixed Communities of Humans, Bears, Sheep, and Wolves'. *Inquiry* 22, pp. 231–41.
Næss, A. (1989): 'From Ecology to Ecosophy, from Science to Wisdom'. *World Futures* 27, pp. 185–90.
Næss, A. (1989): *Ecology, Community and Lifestyle*. Cambridge: Cambridge University Press.
Næss, A. (2005): *The Selected Works of Arne Naess*. Edited by Alan Drengson in cooperation with Arne Næss. (Ten volumes, of which Vol. 10 is the one featuring the most explicit ecophilosophical texts included in the selection.) Dordrecht: Springer.
Næss, A. (2008): *Life's Philosophy. Reason & Feeling in a Deeper World*. London and Athens, Georgia: The University of Georgia Press.
Næss, A. (2008): *The Ecology of Wisdom. Writings by Arne Naess*. Edited by Alan Drengson and Bill Dewall. Berkeley, California: Counterpoint.

Georg Henrik von Wright

Von Wright, G. (1993): *The Tree of Knowledge and Other Essays*. Leiden: Brill. (Especially the essays in the second of the book's two parts are relevant in an ecophilosophical context.)

INDEX